Bismuth - Fundamentals and Optoelectronic Applications

Edited by Yanhua Luo, Jianxiang Wen and Jianzhong Zhang

Published in London, United Kingdom

IntechOpen

Supporting open minds since 2005

Bismuth - Fundamentals and Optoelectronic Applications
http://dx.doi.org/10.5772/intechopen.80205
Edited by Yanhua Luo, Jianxiang Wen and Jianzhong Zhang

Contributors
Alexander Kir'Yanov, Arindam Halder, Yanhua Luo, Bowen Zhang, Mingjie Ding, Shuen Wei, BinBin Yan, Gang-Ding Peng, Rifat M. Dakhil Alsingery, A. Mudhafer, Jianxiang Wen, Ying Wan, Yanhua Dong, Yi Huang, Fufei Pang, Tingyun Wang, Khursheed Ahmad, Jagadeesh Angadi V, Kuldeep Chand Verma, Jianzhong Zhang, H.R. Lakshmiprasanna, K. Manjunatha

First published in London, United Kingdom, 2020 by IntechOpen
IntechOpen is the global imprint of INTECHOPEN LIMITED, registered in England and Wales, registration number: 11086078, 5 Princes Gate Court, London, SW7 2QJ, United Kingdom
Printed in Croatia

British Library Cataloguing-in-Publication Data
A catalogue record for this book is available from the British Library

Additional hard and PDF copies can be obtained from orders@intechopen.com

Bismuth - Fundamentals and Optoelectronic Applications
Edited by Yanhua Luo, Jianxiang Wen and Jianzhong Zhang
p. cm.
Print ISBN 978-1-83968-242-1
Online ISBN 978-1-83968-243-8
eBook (PDF) ISBN 978-1-83968-244-5

For EU product safety concerns:
IN TECH d.o.o., Prolaz Marije Krucifikse Kozulić 3, 51000 Rijeka, Croatia, info@intechopen.com or visit our website at intechopen.com.

Meet the editors

Dr. Yanhua Luo received his B.E. and PhD degrees from the University of Science and Technology of China (UTSC) in 2004 and 2009, respectively. Currently, he works as a deputy director of Photonics & Optical Communications at the University of New South Wales (UNSW) to maintain the National Joint Fibre Facility and develop the next generation specialty optical fibers and their devices. He has made many contributions, i.e. developing 3 modes EDFA for mode multiplexing transmission by NEC America, introducing 3D printing techniques in fiber fabrication, etc. So far he has held 2 China patents and co-authored 220 refereed journal/conference papers, and 7 book chapters on these subjects. His papers have been cited >1500 times in SCI, with an H-index 21.

Prof. Jianxiang Wen received his M.S. degree in chemical and material engineering from Jiangnan University, Jiangsu, China, in 2006, and received his PhD in communication and information systems from Shanghai University, Shanghai, China in 2011. From 2001 to 2007, he was an R&D Engineer in the Jiangsu Fasten Photonics Company. Now he is a Professor and works at the Key Laboratory of Specialty Fiber Optics and Optical Access Networks, Shanghai University. His research interests include design and fabrication of the specialty fiber fields such as spun optical fibers, polarization-maintaining optical fibers, doping fibers, specialty optical fibers with radiation-hardness, etc. His awards include Jiangsu Province Science and Technology Progress Award (Second Prize) in 2008.

Prof. Jianzhong Zhang received his Bachelor degree of condensed-state physics from the Lanzhou University in 2000 and obtained his Master and Doctoral degrees in optical engineering from the Harbin Engineering University in 2004 and 2007, respectively. He then joined the School of Physics at Harbin Engineering University. Currently, he is a full professor of Harbin Engineering University. During 2006 he visited UNSW as a visiting fellow supervised by Professor Gang-Ding Peng. His research interests are in optical fiber laser, optical fiber sensors and wave characteristics in periodical structure. He has published more than 100 articles in international journals and conferences. He is currently the PI for six research projects including two funded by the National Science Foundation of China.

Contents

Preface

The development of the information society drives the increasing demand for novel optoelectronic materials. Therefore, bismuth-related optoelectronic materials are increasingly favoured due to the unique characteristics of bismuth. This book mainly concentrates on the development of optoelectronic materials for the applications of photon generation, amplification, detection and storage, although there is still a lot of outstanding and innovative research work in this field.

The book contains 8 chapters contributed to by many excellent researchers. It is organized in the following sections:

1. Introduction

2. Bismuth Optical Fibers

3. Bismuth Halide Perovskite Solar Cells

4. Bismuth-Related Data Storage Materials

Several sections and chapters of the book show how diverse bismuth-related optoelectronic materials are becoming. It is expected that, in the near future, many new bismuth-related optoelectronic materials under development will find important applications in telecommunications, renewable energy, data storage, and so on. All the authors are leading researchers in their respective fields. Their chapters reflect the excellent research work and technology applications of their own work and others. Hence this book, as a new entry to the open access IntechOpen book library, will be useful for researchers, academics, engineers, and students to access expertly summarized specific topics on bismuth-related optoelectronic materials for research, education, and learning purposes.

We would like to take this opportunity to express our great appreciation to all our colleagues and authors as well as collaborators for their support and great contribution to this book. We also would like to thank the IntechOpen editors and staff, especially Mr. Mateo Pulko, Mr. Luka Cvjetkovic, Ms. Marina Dusevic for their excellent support throughout this book project.

Yanhua Luo
Photonics and Optical Communications,
School of Electrical Engineering and Telecommunications,
University of New South Wales,
Sydney, Australia

Jianxiang Wen
Key Laboratory of Specialty Fiber Optics and Optical Access Networks,
Joint International Research Laboratory of Specialty Fiber Optics
and Advanced Communication,
Shanghai University,
Shanghai, China

Jianzhong Zhang
Key Laboratory of In-Fiber Integrated Optics,
Ministry Education of China,
Harbin Engineering University,
Harbin, China

Section 1

Introduction

Chapter 1

Introductory Chapter: Bismuth-Related Optoelectronic Materials

Yanhua Luo, Jianxiang Wen and Jianzhong Zhang

1. Introduction

21st century is the information era. In this rapidly developing information society, more and more optoelectronic materials are needed to meet the increasing demands, including the information generation, transmission, amplification, detection, storage, etc. Among many optoelectronic materials, Bi-related optoelectronic materials are an essential category. The unique properties of bismuth enable the diversity of functions and applications. As schemed in **Figure 1**, bismuth related optoelectronic materials have demonstrated the great potential for the photon generation, amplification, detection and storage. For example, Bi-doped optical fibers (BDFs) have already been proved as the promising active media for the creation of BDF lasers and amplifiers in the near infrared (NIR) region from 1150 to 1800 nm, including the regions of 1250 – 1500 nm and 1600-1800 nm, where efficient rare-earth fiber lasers are absent [1]. In addition, Bi/Er and Bi/Er/Yb co-doped optical fiber (BEDF and BEYDF) have further been developed for broader bandwidth [2, 3]. The BDF lasers can cover O, E, S, C, L and U bands, which are commonly used for fiber-optic telecommunication [4, 5]. The optical amplification in O, E, S, C, L and U bands have also been achieved in both BDFs and BEDFs [6–8]. Regarding to the commercialized bismuth-doped fiber amplifier (BDFA), OFS company reported that the BDFA operating over the O-band (1272-1310 nm) extend 425 Gb/s 400GBASE-LR8 transmission beyond 50 km of G.652 fiber [8]. In the detection field, high performance of the broadband photo detecting have been realized in many bismuth related materials, like bismuth selenide nanowire [9], bismuth telluride nanoplate [10], bismuth sulfide nanobelt [11], bismuth nanosheet [12], and bismuth film [13]. In addition, bismuth halide perovskites have been demonstrated as one of the most efficient and promising solar cells for its free of toxic Pb [14]. Bismuth thin film [15], bismuth tellurite [16], bismuth-substituted iron garnet [17], etc. have exhibited great potential for data storage. In addition, both Bi^{3+} substituted spinel nanoparticles [18] and multiferroic $BiFeO_3$ [19] can be used for the data storage.

The applications of bismuth related optoelectronic materials above are mainly derived from the unique characteristics of bismuth. It is well known that bismuth is a post-transition metal element. Bismuth and its ions have multivalent states ranged from +5 to 0, or even negative valence [20]. Its ions belong to 6p ions, whose valence state is easily changed, particularly under high temperature and reduction atmosphere [21]. Generally, Bi oxides are likely amphoteric (acidic - basic) for lower valence or acidic for higher valence [22]. As Bi^{82} and Pb^{83} are located nearby with each other on the periodic table, they have many isoelectronic configurations, e.g. Pb and Bi^{+}, Pb^{+} and Bi^{2+}, and Pb^{2+} and Bi^{3+}. Isoelectronic configurations not

Figure 1.
The functionality of bismuth related optoelectronic materials.

only allow the replacement of Pb with Bi, but also make their properties similar, including density of their doped borate glasses [23], capability to generate the non-bridging oxygens [24], diamagnetic property [25], the analogy of the photoluminescence [21], etc.

Although there are many reports of bismuth related optoelectronic materials at present, this book mainly focuses on their applications for the photon generation, amplification, detection and storage. The whole book can be structured into three sections:

Section 1: Since the first demonstration of broad NIR luminescence in Bi doped silica glass in 1999 [26], BDF and BEDF has been developed for the full utilization of the huge unused bandwidth of the existing telecommunication network, in response to the forecasted upcoming 'capacity crunch'. In this section, the development of BDF and the post-treatment effect by radiations, like laser, gamma ray and electron beam have been described. Firstly, R. M. D. Alsingery et al. gave an overview of the development of BDFs and their applications in the optical communication system. Then, B. Zhang et al. systematically summarized the photobleaching effect of bismuth active centers (BACs) related to the NIR luminescence in BDF and BEDF, in terms of irradiation intensity/wavelength and temperature. Subsequently, J. Wen et al. fabricated Bi co-doped silica optical fibers with atomic layer deposition (ALD) and modified chemical vapor deposition (MCVD). In addition, gamma radiation effect upon the fluorescence intensity/lifetime, the absorption, as well as Verdet constants of BDFs fabricated have been investigated. Finally, A. Kir'yanov et al. studied effects of electron irradiation on optical properties of bismuth doped phosphosilicate fiber. The results reveal its overall resistance to irradiation in terms of emission and bleaching contrast at excitation into the absorption bands of BACs. In addition, they found a new effect of large dose-dependent Stokes shift, experienced by the fiber's cutoff wavelength due to the radiation induced refractive-index rise in its core area. The studies of these researchers not only give more information about the configuration and structure of BACs, but also offer an effective approach to regulate BACs.

Section 2: In response to the energy crisis, more and more photovoltaic materials are developed. In recent years, the perovskite solar cells were developed with an excellent power conversion efficiency of 25%, which has been considered as one of the most efficient and promising solar cells. To overcome the remaining issues, like the presence of highly toxic lead and poor stability, bismuth has been introduced to replace Pb in perovskite solar cells due to their similar property. In this section, K. Ahmad overviewed the fabrication of bismuth halide perovskite solar cells. At the end, some strategies for the improved performance are prospected.

Section 3: In recent years, with the rapid development of optical fiber communication networks, Internet of Things, big data and cloud computing, the amount of global data has shown the exponential growth. The global data volume has already exceeded 45 ZB in 2019, and it is expected to reach 175 ZB in 2025 [27]. To meet the ever-increasing data storage, it is very important to develop the new generation of data storage media for computing performance and the magnetic random access memories storage. In this section, K. C. Verma and J. Angadi described the synthesis and characterization of two bismuth related data storage materials: multiferroic $BiFeO_3$ and Bi^{3+} doped manganese spinel ferrite nanoparticles, respectively.

The ferroelectricity of $BiFeO_3$ is originated from $6s^2$ lone-pair electrons of Bi^{3+} and the G-type antiferromagnetic ordering resulted from Fe^{3+} spins order of cycloidal below Neel temperature. So multiferroic $BiFeO_3$ allows the data to be written electrically and read magnetically due to its magnetoelectric effect observed at room temperature. Especially, the structure of $BiFeO_3$ can be controlled through the selection of appropriate synthesis route, reaction conditions and heating processes. To overcome the drawback of the disappearance of multiferroicity in $BiFeO_3$, several solutions are proposed like the replacement of dopant ions, the control of ions concentration, $BiFeO_3$ composites as well as thin films, especially multilayer structures.

The $Mn_{1-x}Bi_xFe_2O_4$ ($x = 0.0, 0.05, 0.1, 0.15$ and 0.2) nanoparticles were prepared by the solution combustion method and their structural, microstructural and magnetic properties are characterized. Rietveld refined X-ray diffraction (XRD) patterns confirm the single-phase formation with space group Fd3m having spinel cubic structure and the lattice parameters increase with the increase of Bi^{3+} concentration. The magnetic hysteresis curves reveal the soft magnetic material nature of these nanoparticles, demonstrating the great potential for absorbing electromagnetic waves, storage media, etc.

As the most promising gain medium for the next generation photonic network, the work presented in the Section 1 is evidently not enough. So, challenges and opportunities of the development of BDF and BEDF is further briefed in the following.

2. Challenges and opportunities of bismuth doped optical fiber

2.1 Background

Water-free fiber technology has enabled silica optical fibers to achieve ultra-low loss transmission over the broad spectral range of 1200-1700 nm, where the transmission loss is less than 0.5 dB/km as shown in **Figure 2** [28]. **Figure 2** also shows the NIR spectral range of some rare earth ions often doped in fibers [28]. It has already demonstrated that the construction of the efficient optical amplifiers based on rare earths doped optical fibers for the extended bands is impossible. Though many researchers have previously been focused on the development of fiber amplifiers and lasers based on fibers doped with rare earth ions, e. g. Er^{3+}, Tm^{3+}, Yb^{3+}, Nd^{3+}, Ho^{3+}, Pr^{3+}, for the extended bands, none of them has demonstrated sufficient optical gain and enough bandwidth for the telecommunication applications.

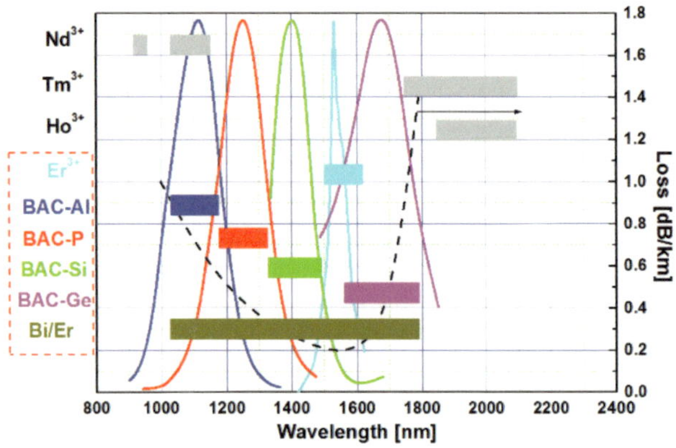

Figure 2.
Spectral ranges of various doping elements, normalized emission spectra from each active center interested as well as the low-loss spectrum of silica-based optical fiber.

Since the first demonstration of broadband NIR luminescence in Bi-doped silica glass [26], many researchers have devoted to the investigation of bismuth doped glasses or materials for the extended band [29–31]. The development in these materials has spurred significant work in optical fiber form [28, 32], since the fabrication of the first BDF in 2005 [33]. Different from the well shielded shell of rare earth ions, the electronic shell of bismuth is easy coupled with the external environment, which finally influences the characteristic properties of BACs. According to the local environments, there exist four types of BACs, which are BAC-Si, BAC-Ge, BAC-P and BAC-Al, relating to SiO_2, GeO_2, P_2O_5:SiO_2 and Al_2O_3:SiO_2, respectively [5]. The normalized luminescence spectra of these four BACs are plotted in **Figure 2** [34]. Seen from **Figure 2**, BDFs have demonstrated as the promising active media for the NIR region from 1150 to 1800 nm [1]. Especially, through their co-doping, more broader and stronger emission and gain can be achieved, like Bi/Er [7, 35–37], Bi/Tm [38, 39], etc. [40, 41].

2.2 Remaining challenges

Despite the great success in the development of the BDF technology for the unused bandwidth, there remain many fundamental scientific and technological issues and challenges, before these fiber lasers and amplifiers can be practically and commercially used.

On the scientific side: The main challenge is that the nature of bismuth NIR emitting centers is still not clear [1]. Difficulties mainly arise from Bi where its *d* orbitals are easily coupled to the surrounding environment. But it is such coupling, which allows the formation of broadband NIR emission center through the tailoring of the external environment by additional dopants such as Ge, P, Al, etc. Meanwhile, bismuth as the wonder metal, can proceed reduction reactions with no other element and produce such a variety of products [42]. Especially, under the high temperature, the change of the valent state of bismuth in Bi-doped glasses occurs [43]: $Bi^{3+} \rightarrow Bi^{2+} \rightarrow Bi^{+} \rightarrow Bi/Bi_2, Bi_2^-, Bi_3$, etc. $\rightarrow (Bi)_n$, where Bi_2, Bi_2^-, Bi_3, etc. are Bi clusters and $(Bi)_n$ is metallic colloid. As bismuth doped glasses and fibers often undergo the high temperature in the fabrication process. For example, under the high temperature treatment, the color of the doped core in the BEDF preform is clearly changed as indicated in **Figure 3** (the dark color is often taken as the evidence of the formation of $(Bi)_n$). So many types of bismuth often co-exist in these Bi doped glasses and fibers. Their properties, like unpaired electrons, radius and fluorescence lifetime are different with each other as listed in **Table 1**.

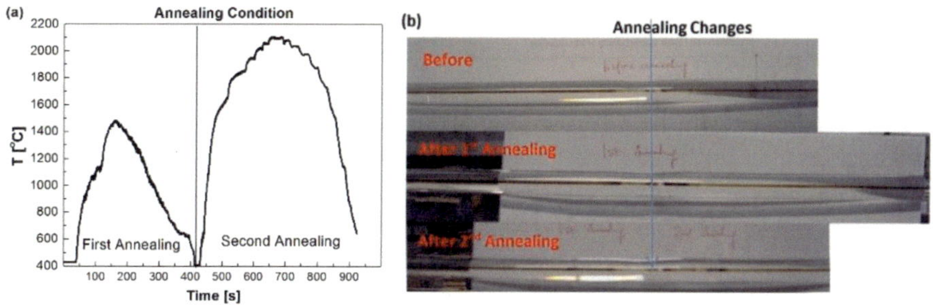

Figure 3.
Thermal treatment influence upon the fiber core in the BEDF preform: (a) annealing conditions; (b) the color change in the core part with thermal treatment.

Valence	Configuration	Number of unpair electrons [44]	Radius [Å] [45]	Lifetime [μs] [46]	
				NIR	Visi
Bi^{0+}	$[Xe]5d^{10}6s^26p^3$	3	1.56	$\geq 10^2$	~10
Bi^{1+}	$[Xe]5d^{10}6s^26p^2$	2	1.45	$\sim 10^2$	≥10
Bi^{2+}	$[Xe]5d^{10}6s^26p^1$	1	1.16	—	~10
Bi^{3+}	$[Xe]5d^{10}6s^2$	0	0.96	—	1
Bi^{4+}	$[Xe]5d^{10}6s^1$	1	—	—	—
Bi^{5+}	$[Xe]5d^{10}$	0	0.74	—	—
Bi^{6+}	$[Xe]5d^9$	1	—	—	—
BiO	—	—	1.70	$\geq 10^2$	≥10
Bi_2^-	—	—	—	$\geq 10^2$	~10

Table 1.
Upaired electrons, radius and fluorescence lifetime of bismuth ions.

So far, a number of hypotheses are reported about the origin of the NIR emission center in Bi-doped glasses, like: Bi^+, Bi^{5+}, Bi^-clusters, BiO, Bi_2^-, Bi_2^{2-} point defects and others, but none of them have been directly confirmed [1]. Though the recent report by Dianov et al. has confirmed that Bi-related NIR emitting centers are clusters consisting of Bi ions and oxygen deficiency centers instead of Bi ions themselves [1], the exact form of bismuth in BAC remains unclear. In addition, the NIR luminescence has also been observed from Bi-doped CsI halide crystals [47], fluoride glasses [48], etc. where no oxygen deficiency will exist in these materials. Therefore, to conquer this challenge, further research is necessary to get reliable results on the nature of Bi-related NIR emitting centers in Bi-doped glasses and optical fibers of various compositions and fabrication techniques with different processing conditions.

On the technological side: As the valence state of Bi in BDF/BEDF will change during the fabrication process, so it becomes a technical challenge to control the formation rate of BAC for the preparation of optical fibers. Both the valence state of bismuth and oxygen deficiency centers are sensitive to the processing temperature [43, 49]. In addition, bismuth oxide is easier evaporated during the collapsing process of preform fabrication with the MCVD technique compared with rare earth oxide. Hence, the control of the valence state of Bi as well as the formation rate of BAC is hardly achieved in BDF/BEDF. So there often co-exist many types of Bi ions or BAC in Bi-doped materials [50]. Though the full control of the formation of

BACs is very hard, reductive agents, like high-purity silicon powder or sucrose in MCVD process [51] and SiC in melt and pour glass [52] have already been introduced in the fabrication of the Bi-doped materials as bismuth NIR emitting centers are formed in an endothermic redox chemical reaction [53]. In addition, post treatments, like femtosecond laser [54], thermal treatment [55, 56], γ-radiation [57], H_2 reduction [58], etc. have also been tried to activate and control the BAC. Recently, M. Melkumov et al. have tried to improve the performance of BDF by the optimization of drawing and MCVD processing conditions [59, 60]. Though these solutions can regulate the formation of BAC to some degree, the success rate still cannot be quantified due to the unclear structure of the BAC. In addition, most of BDFs with lasing and amplification are doped with very low content of bismuth (usually <0.1 wt.%) [5], which is often lower than the detection limit of mostly used equipment-energy-dispersive X-ray (EDX) analyzer. However, the performance improvement by the increment of Bi concentration is limited by the fast growth of the background loss in fiber [5]. So the concentration increment and the control growth of the background loss should be balanced for high performance BDF and BEDF. All these technical issues are subject to further in-depth and systematic research and development, such as the selection of dopants and their compositions, and fabrication conditions, etc.

Author details

Yanhua Luo[1]*, Jianxiang Wen[2] and Jianzhong Zhang[3]

1 Photonics and Optical Communications, School of Electrical Engineering, University of New South Wales, Sydney 2052, NSW, Australia

2 Key Laboratory of Specialty Fiber Optics and Optical Access Networks, Joint International Research Laboratory of Specialty Fiber Optics and Advanced Communication, Shanghai Institute for Advanced Communication and Data Science, Shanghai University, Shanghai 200444, China

3 Key Laboratory of In-Fiber Integrated Optics of Ministry of Education, School of Science, Harbin Engineering University, Harbin 150001, China

*Address all correspondence to: yanhua.luo1@unsw.edu.au

References

[1] Dianov EM, Nature of Bi-related near IR active centers in glasses: state of the art and first reliable results, Laser Physics Letters, **12**(9), 095106, 2015

[2] Luo Y, Wen J, Zhang J, Canning J, and Peng G-D, Bismuth and erbium codoped optical fiber with ultrabroadband luminescence across O-, E-, S-, C-, and L-bands, Optics Letters, **37**(16), 3447-3449, 2012

[3] Sathi ZM, Zhang J, Luo Y, Canning J, and Peng GD, Improving broadband emission within Bi/Er doped silicate fibres with Yb co-doping, Optical Materials Express, **5**(10), 2096-2105, 2015

[4] Firstov S, Alyshev S, Melkumov M, Riumkin K, Shubin A, and Dianov E, Bismuth-doped optical fibers and fiber lasers for a spectral region of 1600-1800 nm, Optics Letters, **39**(24), 6927-6930, 2014

[5] Bufetov IA, Melkumov MA, Firstov S V, Riumkin KE, Shubin AV, Khopin VF et al., "Bi-doped optical fibers and fiber lasers," IEEE Journal of Selected Topics in Quantum Electronics, **20**(5), 0903815, 2014

[6] Dvoyrin VV, Medvedkov OI, MashinskyVM,UmnikovAA,GuryanovAN, and Dianov EM, "Optical amplification in 1430-1495 nm range and laser action in Bi-doped fibers," Optics Express, **16**(21), 16971-16976, 2008

[7] Firstov SV, Firstov SV, Riumkin KE, Khegai AM, Alyshev SV, Melkumov MA, Khopin VF et al., Wideband bismuth- and erbium-codoped optical fiber amplifier for C + L + U-telecommunication band, Laser Physics Letters, **14**(11), 110001, 2017

[8] Mikhailov V, Melkumov MA, Inniss D, Khegai AM, Riumkin KE, Firstov SV et al., Simple broadband bismuth doped fiber amplifier (BDFA) to extend O-band transmission reach and capacity, in Optical Fiber Communication Conference (OFC) 2019, San Diego, California, 2019: Optical Society of America, p. M1J.4

[9] Sharma A, Bhattacharyya B, Srivastava AK, Senguttuvan TD, and Husale S, High performance broadband photodetector using fabricated nanowires of bismuth selenide, Scientific Reports, **6**(1), 19138, 2016

[10] Zhang Y, You Q, Huang W, Hu L, Ju J, Ge Y et al. Few-layer hexagonal bismuth telluride (Bi_2Te_3) nanoplates with high-performance UV-Vis photodetection, Nanoscale Advances, **2**(3), 1333-1339, 2020

[11] Xu J, Li H, Jiang K, Yao H, Fang F, Chen F et al., Synthesis of bismuth sulfide nanobelts for high performance broadband photodetectors, Journal of Materials Chemistry C, **8**(6), 2102-2108, 2020

[12] Huang H, Ren X, Li Z, Wang H, Huang Z, Qiao H et al., Two-dimensional bismuth nanosheets as prospective photo-detector with tunable optoelectronic performance, Nanotechnology, **29**(23), 235201, 2018

[13] Yao JD, Shao JM, and Yang GW, Ultra-broadband and high-responsive photodetectors based on bismuth film at room temperature, Scientific Reports, **5**(1), 12320, 2015

[14] Ahmad K, Ansari S.N, Natarajan K, and Mobin SM, Design and synthesis of 1D-polymeric chain based $[(CH_3NH_3)_3Bi_2Cl_9]_n$ perovskite: a new light absorber material for lead free perovskite solar cells, ACS Applied Energy Materials, **1**(6), 2405-2409, 2018

[15] Suh S and Craighead H, Optical writing characteristics of multilayered bismuth/selenium thin films, Optical Engineering, **26**(6), 266524, 1987

[16] Oliveira ID, Carvalho JF, Fabris ZV, and Frejlich J, Holographic recording and characterization of photorefractive Bi_2TeO_5 crystals at 633 nm wavelength light, Journal of Applied Physics, **115**(16), 163514, 2014

[17] Anoikin EV and Sides PJ, "Plasma-activated chemical vapor deposition of bismuth-substituted iron garnets for magneto-optical data storage," IEEE Transactions on Magnetics, **31**(6), 3239-3241, 1995

[18] Seevakan K, Manikandan A, Devendran P, Slimani Y, Baykal A, and Alagesan T, Structural, magnetic and electrochemical characterizations of $Bi_2Mo_2O_9$ nanoparticle for supercapacitor application, Journal of Magnetism and Magnetic Materials, **486**, 165254, 2019

[19] Wang J, Neaton JB, Zheng H, Nagarajan V, Ogale SB, Liu B et al., Epitaxial $BiFeO_3$ multiferroic thin film heterostructures, Science, **299**(5613), 1719-1722, 2003

[20] Xu L, Bobev S, El-Bahraoui J, and Sevov SC, A naked diatomic molecule of bismuth, $[Bi_2]^{2-}$, with a short Bi–Bi bond: synthesis and structure, Journal of the American Chemical Society, **122**(8), 1838-1839, 2000

[21] Hughes MA, Gwilliam RM, Homewood K, Gholipour B, Hewak DW, Lee T-H et al., On the analogy between photoluminescence and carrier-type reversal in Bi- and Pb-doped glasses, Optics Express, **21**(7), 8101-8115, 2013

[22] http://ww2.chemistry.gatech.edu/~barefield/1311/coordination_complexes.pdf (accessed 20/09/2020)

[23] Khanisanij M and Sidek HAA, Elastic behavior of borate glasses containing lead and bismuth oxides, Advances in Materials Science and Engineering, **2014**, 452830, 2014

[24] Sidek HAA, Hamezan M, Zaidan AW, Talib ZA, and Kaida K, Optical characterization of lead-bismuth phosphate glasses, American Journal of Applied Sciences, **2**(8), 1266-1269, 2005.

[25] Elisa M, Iordanescu R, Vasiliu C, Sava BA, Boroica L, Valeanu M, Kuncse V et al., Magnetic and magneto-optical properties of Bi and Pb-containing aluminophosphate glass, Journal of Non-Crystalline Solids, **465**, 55-58, 2017

[26] Murata K, Fujimoto Y, Kanabe T, Fujita H, and Nakatsuka M, Bi-doped SiO_2 as a new laser material for an intense laser, Fusion Engineering and Design, **44**(1), 437-439, 1999

[27] Reinsel D, Gantz J, and Rydning J, "The digitization of the world from edge to core," IDC, 2018.

[28] Dianov EM, Amplification in extended transmission bands using bismuth-doped optical fibers, Journal of Lightwave Technology, **31**(4), 681-688, 2013

[29] Fujimoto Y and Nakatsuka M, Infrared luminescence from bismuth-doped silica glass, Japanese Journal of Applied Physics, **49**(2001), L279-L281, 2001

[30] Meng X, Qiu J, Peng M, Chen D, Zhao Q, Jiang X et al., Near infrared broadband emission of bismuth-doped aluminophosphate glass, Optics Express, **13**, 1628-1634, 2005

[31] Sun H-T, Zhou J, and Qiu J, Recent advances in bismuth activated photonic materials, Progress in Materials Science, **64**, 1-72, 2014

[32] Dianov EM, Bismuth-doped optical fibres: A new breakthrough in near-IR lasing media, Quantum Electronics, **42**(9), 754 – 761, 2012

[33] Dvoyrin VV, Mashinsky VM, Dianov EM, Umnikov AA, Yashkov MV, and Guranov AN, Absorption, fluorescence and optical amplification in MCVD Bismuth-doped silica glass optical fibres, ECOC Proceeding, **4**, 949-950, 2005

[34] Dianov EM, Firstov SV, and Melkumov M A, Bismuth-doped optical fibers: advances and bew developments, in Workshop on Specialty Optical Fibers and Their Applications, Hong Kong, 2015: Optical Society of America, in OSA Technical Digest (online), p. WT1A.4

[35] Kuwada Y, Fujimoto Y, and Nakatsuka M, Ultrawideband light emission from bismuth and erbium doped silica, Japanese Journal of Applied Physics, **46**, 1531-1532 2007

[36] Peng M, Zhang N, Wondraczek L, Qiu J, Yang Z, and Zhang Q, Ultrabroad NIR luminescence and energy transfer in Bi and Er/Bi co-doped germanate glasses, Optics Express, **19**(21), 20799-20807, 2011

[37] Luo Y, Wen J, Zhang J, Canning J, and Peng G-D, Bismuth and erbium codoped optical fiber with ultrabroadband luminescence across O-, E-, S-, C-, and L-bands, Optics Letters, **37**(16), 3447-3449, 2012

[38] Ruan J, Dong G, Liu X, Zhang Q, Chen D, and Qiu J, Enhanced broadband near-infrared emission and energy transfer in Bi–Tm-codoped germanate glasses for broadband optical amplification, Optics Letters, **34**(16), 2486-2488, 2009

[39] Zhou B, Lin H, Chen B, and Pun EY-B, Superbroadband near-infrared emission in Tm-Bi codoped sodium-germanium-gallate glasses, Optics Express, **19**(7), 6514-6523, 2011

[40] Ruan J, Chi Y, Liu X, Dong G, Lin G, Chen D et al., Enhanced near-infrared emission and broadband optical amplification in Yb–Bi co-doped germanosilicate glasses, Journal of Physics D: Applied Physics, **42**(15), 155102, 2009

[41] Zhang P, Chen N, Wang R, Huang X, Zhu S, Li Z et al., Charge compensation effects of Yb^{3+} on the Bi^{+}: near-infrared emission in PbF_2 crystal, Optics Letters, **43**(10), 2372-2375, 2018

[42] Corbett JD, Homopolyatomic ions of the post-transition elements—synthesis, structure and bonding (Progress in Inorganic Chemistry). New York: Wiley, 1976

[43] Khonthon S, Morimoto S, Arai Y, and Ohishi Y, Redox equilibrium and NIR luminescence of Bi_2O_3-containing glasses, Optical Materials, **31**(8), 1262-1268, 2009

[44] Ohkura T, Fujimoto Y, Nakatsuka M, and Young-Seok S, Local structures of bismuth ion in bismuth-doped silica glasses analyzed using Bi L_{III} X-ray absorption fine structure, Journal of the American Ceramic Society, **90**(11), 3596-3600, 2007

[45] Sun H-T, Matsushita Y, Sakka Y, Shirahata N, Tanaka M, Katsuya Y et al., Synchrotron X-ray, photoluminescence, and quantum chemistry studies of bismuth-embedded dehydrated zeolite Y, Journal of the American Chemical Society, **134**(6), 2918-2921, 2012

[46] Sokolov VO, Plotnichenko VG, and Dianov EM, The origin of near-IR luminescence in bismuth-doped silica and germania glasses free of other dopants: First-principle study, Optical Materals Express, **3**(8), 1059-1074, 2013

[47] Su L, Zhao H, Li H, Zheng L, Ren G, Xu J et al., Near-infrared ultrabroadband luminescence spectra properties of subvalent bismuth in CsI halide crystals, Optics Letters, **36**(23), 4551-4553, 2011

[48] Romanov AN, Haula EV, Fattakhova ZT, Veber AA, Tsvetkov VB, Zhigunov DM et al., Near-IR luminescence from subvalent bismuth species in fluoride glass, Optical Materials, **34**(1), 155-158, 2011

[49] Hanafusa H, Hibino Y, and Yamamoto F, Formation mechanism of drawing-induced defects in optical fibers, Journal of Non-Crystalline Solids, **95-96**, 655-661, 1987

[50] Xu W, Peng M, Ma Z, Dong G, and Qiu J, A new study on bismuth doped oxide glasses, Optics Express, **20**(14), 15692-15702, 2012

[51] Li J, Jiang Z, Peng J, Dai N, Li H, Yang L et al. Preparation method of bismuth-doped silica fiber controllable in components and valence state, and bismuth-doped silica fiber, China Patent, CN103601364A, 2020

[52] Cao J, Xu S, Zhang Q, Yang Z, and Peng M, Ultrabroad photoemission from an amorphous solid by topochemical reduction, Advanced Optical Materials, **6**(22), 1801059, 2018

[53] Veber A, Cicconi MR, Puri A, and de Ligny D, Optical properties and bismuth redox in Bi-doped high-silica Al–Si glasses, The Journal of Physical Chemistry C, **122**(34), 19777-19792, 2018

[54] Kononenko V, Pashinin V, Galagan B, Sverchkov S, Denker B, Konov V, et al., Laser induced rise of luminescence efficiency in Bi-doped glass, Physics Procedia, **12**, 156-163, 2011

[55] Wei S, Luo Y, Fan D, Xiao G, Chu Y, Zhang B et al., BAC activation by thermal quenching in bismuth/erbium codoped fiber, Optics Letters, 44(7), 1872-1875, 2019

[56] Kharakhordin AV, Alyshev SV, Firstova EG, Lobanov AS, Khopin VF, Khegai AM et al., Lasing properties of thermally treated GeO_2-SiO_2 glass fibers doped with bismuth," Applied Physics B, **126**(5), 87, 2020

[57] D. Sporea, Mihai L, Neguț D, Luo Y, Yan B, Ding M et al., γ irradiation induced effects on bismuth active centres and related photoluminescence properties of Bi/Er co-doped optical fibres, Scientific Reports, **6**(1), 29827, 2016

[58] Nielsen KH, Smedskjaer MM, Peng M, Yue Y, and Wondraczek L, Surface-luminescence from thermally reduced bismuth-doped sodium aluminosilicate glasses, Journal of Non-Crystalline Solids, **358**(23), 3193-3199, 2012

[59] S. Firstov, Levchenko A, Kharakhordin A, Khegai A, Alyshev S, Melkumov M et al., Effect of drawing conditions on optical properties of bismuth-doped high-GeO_2–SiO_2 fibers, IEEE Photonics Technology Letters, **32**(15), 913-916, 2020

[60] Khegai A, Afanasiev F, Ososkov Y, Riumkin K, Khopin V, Lobanov A et al., The influence of the MCVD process parameters on the optical properties of bismuth-doped phosphosilicate fibers, Journal of Lightwave Technology, doi: 10.1109/JLT.2020.3008536, 2020

Section 2

Bismuth Optical Fibers

Chapter 2

Development of Bismuth-Doped Fibers (BDFs) in Optical Communication Systems

Rifat M. Dakhil Alsingery and Ahmed Mudhafer

Abstract

This chapter will provide background information in the development of BDFs and their applications in optical communication systems. Herein, the main focus is briefly described previous studies on BDFs that have attracted much interest over the last two decades. This necessary information and concepts are very much relevant to understanding this book, mainly due to the doping of Bi in the studied bismuth and erbium-doped silicate fibers (BEDFs). The remaining chapter is consisting of the following sections: Sec.2: General introduction about optical fibers. Sec. 3 discusses the general spectral characteristics of BDFs. Sec.4: Including the active centers (namely the bismuth (Bi) active centers (BACs)) responsible for the spectral properties in Bi-doped fibers. Sec.4 Discusses the Bismuth Doped Fiber Amplifier (BDFA).

Keywords: bismuth-doped fibers (BDFs), optical communication, Bi active centers (BACs), bismuth doped fiber amplifier (BDFA)

1. Introduction

A previous work by C.Kato et al. stated that the loss of 20 dB/km in a dielectric waveguide could be minimized if it uses pure dielectric material [1]. This encourages many researchers to discover methods to minimize failure in optical silica fibers. Since its proper mechanical and optical properties, silica was used as the material for the improvement of optical fibers. A low-loss was established in 1970 by merging with a 20 dB/km loss [2].

Besides, with ultra-pure precursors with impurities of ppb-order (parts per billion), the growth of low-loss (0.2 dB/km) single-mode fibers about 1550 μm is demonstrated to be possible [3]. **Figure 1** displays the attenuating spectrum of conventional single-mode fiber. The wavelength range of 1260–1625 nm is split into several subwavelength bands. A specific ITU-T wavelength range is allocated to each band, as listed in **Table 1**.

After the introduction of low-loss fiber, erbium (Er)-doped (EDFA) technology was introduced in 1987 as an advance to revolutionize optical fiber communication. It allowed transatlantic fiber communication [5]. The progression of fiber development to advance low-loss fiber in conjunction with EDFA technology led to the use of the low-loss range for optical fiber communications from 1530 to 1625 nm (C L band). Throughout the years, several techniques including dense wavelength

1970 by merging with a 20dB / km loss[2].

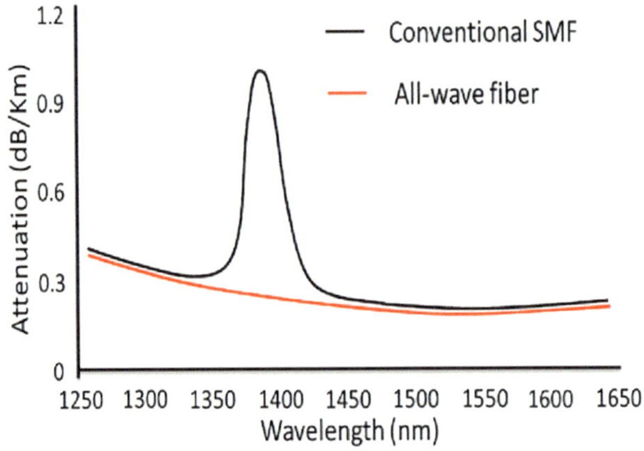

Figure 1.
Attenuation spectra of conventional SMF and all-wave fiber [4].

Bands	O	E	S	C	L
Wavelength (nm)	1260–1360	1360–1460	1460–1530	1530–1565	1565–1625

Table 1.
ITU-T profile of optical wavelength bands in telecommunications.

division multiplexing (DWDM), wavelength multiplexing division (WDM), and coarse wavelength multiplexing division (CWDM) with new modulation formats as quadrature phase-shift keying (QPSK) as 16-QPSK, Quadrature Phase-shift Keying (QPSK) and 64-QPSK used to increase capacities for existing silica optical fibers, in the C + L band. Nonetheless, the stated transmission capacity of Standard Single-Mode Fiber (SMF) achieves the non-linear Shannon limit [6]. **Figure 2** shows growth in the global population's use of the internet every ten years. More than 60% of the world's population is projected to be connected by 2020. The enhanced internet connectivity and comprehensive Internet-based applications such as cloud computing, social media, e-commerce, and e-learning are the drivers of significant demand for fiber transmission. New approaches to improving the potential for existing optical fiber networks must also be considered economically rather than merely adding more traditional single-mode fibers to meet end-user needs. Many researchers have proposed many approaches. One of these approaches is using a low-loss window (1260–1625 nm) of silica optical fibers by improving useful fiber lasers and amplifiers. The other is the development of new fibers such as multi-mode fibers (MMF), multi-core fibers (MCF), and multi-element fibers (MEF) for space-division multiplexing (SDM) in a C band from 1530 to 1565 nm. A further challenge is looking for a different transmission band in a 2000 nm wavelength region with new fibers [6]. The early stages of fiber optic communication in C-bands of about 1550 nm employed silica fibers with a loss of about 0.2 dB / km and a peak of about 1380 nm as shown in **Figure 1**. Recently, Lucent technologies and Optical Fiber Solutions (OFS) have delivered low-loss optical fiber in the entire 1260–1625 nm wavelength range. Such ultra-low-loss fibers are known as dry fibers and give a bandwidth of about 53THz for optical fiber. The mitigation for these ultra-low loss fibers in the 1260–1625 nm wavelength range is less than 0.4 dB/km, as shown in **Figure 1** [4, 8]. Besides, new optical networks have been developed. For nearly every feature, they provide dramatically improved performance over

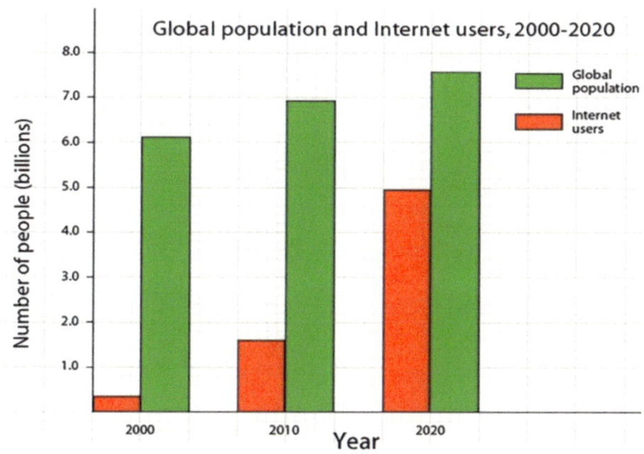

Figure 2.
The bar chart shows the increase in internet users per each ten-year gap [7].

traditional single-mode fiber, including increased usable bandwidth, superior macro-bend performance, and ultra-low dispersion polarization mode (PMD).

10 THz belongs to the Er amplification band between 1530 and 1625 nm with 53THz bandwidth provided by these dry fibers. A bandwidth can, therefore, be increased four times by using dry fibers. All of these dry fiber technologies have provided the ultimate flexibility in network construction and provided economic approaches to optimize data transition capability. However, the industry requires useful fiber amplifiers and lasers to use this utterly low-loss belt for optical communications, which are central to communication between an optical fiber. Regrettably, there are no active rare-earth (RE) doped fiber amplifiers and lasers that support the band of 1260–1530 nm. Given numerous attempts to implement various RE dopers in silica fibers to improve lasers and amplifiers in this wavelength range, potential devices available for use are still lacking. The amplifiers development and lasers in the 1260–1625 nm wavelength band, which uses bismuth (Bi) and Er-doped fibers to use the ultra-low loss window of ultra-low loss optical fiber.

2. Bismuth-doped fibers (BDFs)

Doping of the core zone with RE material is necessary for optical fiber fabricated to improve lasers and amplifiers. Various RE elements and their favorite emission bands in silica host are shown in **Figure 3**. Conventional RE elements, including ytterbium (Yb), Er, and holmium (Ho) or thulium (Tm), cover the wavelength bands about 1000, 1500, and 2000 nm [10]. However, any RE-doped silica does not cover the band between Er and Yb. The Energy level diagram of Yb, Er, Tm is shown in **Figure 4**. neodymium (Nd) and Praseodymium (Pr)-doped fibers explored extensively in silica host to improve lasers and amplifiers about 1300 nm. Nevertheless, highly phonon energy in a silica host made the dopants ineffective. They changed to a low phonon energy host, as fluoride glass enabled them reasonably active. Fluoride glasses, though, are not appropriate for splicing with traditional silica fibers as required in many applications for all-fiber compact devices. Despite this, the fabrication of these fibers is complex and not mature sufficient, in contrast to conventional changed chemical vapor deposition (MCVD)-solution doping technique. Furthermore, as shown in **Figure 4**, in the case of Pr, ground-state absorption (3H4-3F4), excited-state absorption (ESA) (1G4-1D2) at the signal wavelength, and pump ESA (1G4-3P0) are prejudicial to improve effective lasers and Pr-doped fiber amplifiers at 1300 nm.

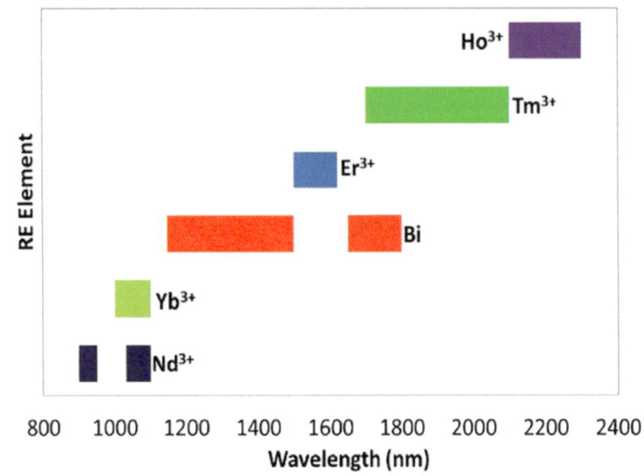

Figure 3.
The Spectrum regions covered via different RE doped elements in silica host [9].

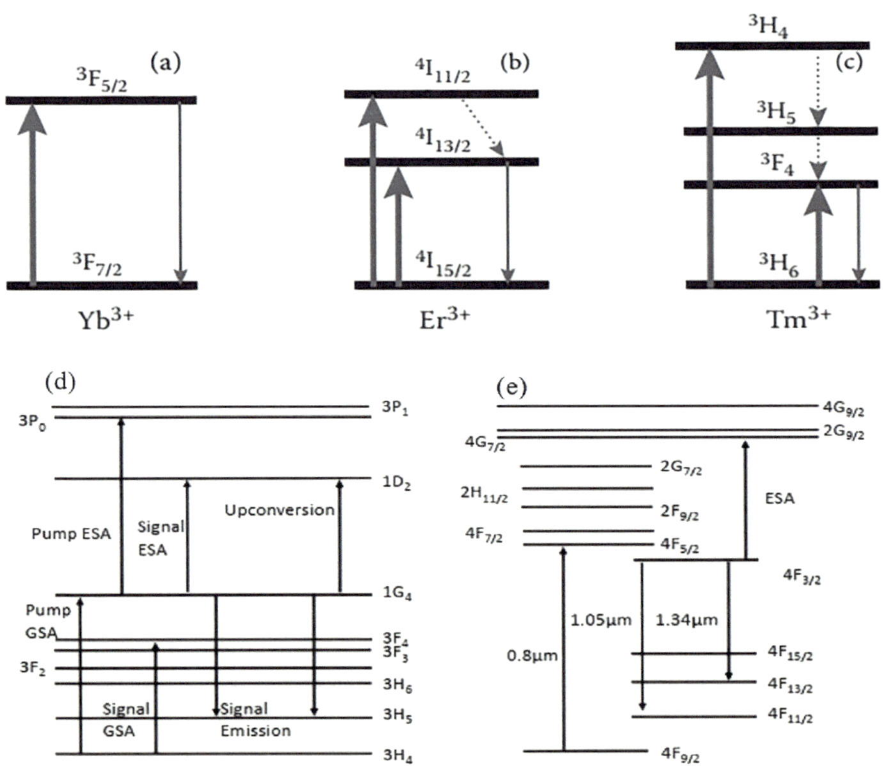

Figure 4.
Energy level diagram of (a) Yb^{3+}, (b) Er^{3+}, (c)Tm^{3+} [10] (d) Pr^{3+} and (e) N^{d3+} [11].

For neodymium (Nd), the ESA at a signal wavelength (4F3/2-4G7/2) with a get competition between 1005 nm (4F3/2-4F11/2) and 1.34 μm (4F3/2-4F13/2) are detrimental to improve 1.3 μm amplifiers and lasers [11]. The last research subjects on BDFs covered the main areas of (i) design of BDFs with new materials and compositions, (ii) studying the fabrication techniques and processes to make BDFs with better properties, (ii) spectral characterization method to assess BDFs for desired applications, (iii) post-fabrication techniques and their impacts on spectral properties, (iv) applications of BDFs into gain media, or sensing device or any other,

(v) evaluation of optical devices made with BDFs using various configurations and operational conditions, (vi) studying the nature of BACs responsible for producing the emission in BDFs, (vii) unraveling the energy levels of the active centers, and many other similar topics. Among these subjects, the nature of BACs is essential for understanding the spectral characteristics and their effects on applications; however, the issue has not been resolved yet and required more addressing.

3. Characteristics and applications of BDFs

3.1 Spectral emission properties and Bi active centers (BACs)

Visible emissions from Bi-doped glasses and crystals were deeply clarified [12–17]. Recently, broadband emissions in the near-infrared (NIR) region from 1200 nm to 1500 nm have been demonstrated in Bi-doped glasses (BDGs) by Fujimoto et al. [18]. Peng also reported broad NIR emission from Bi glasses. et al. as well [19]. With around 1300 nm of optical amplification shown produced by the BDG.s [20]. The first Bi-doped fiber (BDF) and a laser-based on that BDF were reported by Dianov et al. in 2005 [21]. Different BDFs and BDGs have been improved for potential applications as lasers and amplifiers in the second telecommunication window range at around 1.3 μm. It was found that the bandwidth and intensity of emission produced from the BDFs are influenced mainly by the host type and other co-doping materials. Bi can produce emissions in various wavelengths in the NIR wavelength region over (1200–1500) nm, as reported in BDFs and BDGs. For instance, emissions in several BDFs and BDGs (such as silicate, germanate, alumino-silicate, alumino-phosphate, barium-aluminoborate types) have been reported in earlier work [19, 22–27]. Typical emissions of Bi in some recent work are presented in **Figure 5**.

Emissions in BDFs and BDGs are accompanied by active centers or BACs. BACs largely determine the spectroscopic properties, performance, and operations of BDFs. Hence, to design a fiber for a specific application, it is essential to know the BAC details and their relationships to the spectral properties. As an approach to studying the nature of BACs can be broadly categorized into two main groups: (a) spectral analyses to determine the type of BAC and (b) instrumental analyses to identify the chemical bonding and electronic states of Bi, which is responsible for forming a particular type of BAC. The spectral analysis gives information about a specific BAC's spectral characteristics, determined mainly by the composition types in the fiber or glass. Absorption, emission (luminescence or fluorescence), and emission lifetime are some of the most common and basic spectral properties analyzed to identify the BAC types. Besides, an emission-excitation spectroscopic graph described recently was found to be very important to recognize the particular emission [29]. At the initial phase of Bi fiber research, it was thought that Bi emission could be produced only in the presence of Al [23]. However, later, Bi emission was detected in pure silica-doped fibers without any Al co-doping, and lasing was realized in such BDFs [30]. The concept that BACs are linked with other ions in addition to Bi was subsequently considered through the investigation into several Bi fibers and glasses, which were fabricated with different compositions and their relations. Following this advance, Bi fibers with the most straightforward glass compositions were made and BACs in them were characterized, enabling the identification of certain BAC types, e.g., BAC-Si, BAC-Al, BAC-Ge and BAC-P associated respectively with Si, Al, Ge, and P, along with their respective characteristic spectral properties [29–33]. Instrumental analyses of the BAC study concentrate on how to reveal possible valence and the electronic configuration of specific BAC. This approach included analyses of the nuclear magnetic

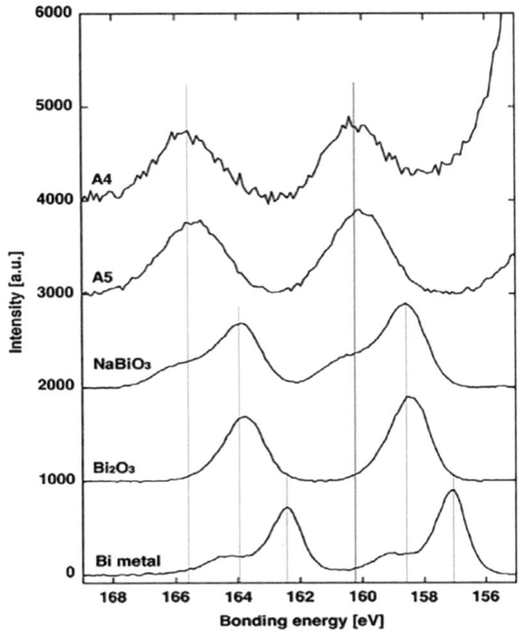

Figure 5.
Characteristic Bi XPS reported by Fujimoto [28].

resonance (NMR), X-ray photoelectron spectroscopy (XPS), electron spin resonance (ESR), X-ray diffraction, and extended X-ray absorption fine structure (EXAFS), *etc.* [28]. Previous research gave rise to different hypotheses about BACs, such as Bi+, Bi0, BiO, Bi clusters [31, 34, 35], and Bi3+ or Bi5+ by Fujimoto *et al.* [28, 32], being responsible for the NIR emissions. Dianov recognized two experimental facts related to BACs: (i) Bi3+ and Bi2+ emit visible emission. However, no NIR emission, and (ii) NIR emissions are observed because of the reduction of Bi3+ and Bi2+ to a lower oxidation state. Therefore, the probable origins of NIR emissions in Bi-doped glass and fiber associated with low valence Bi ions and other dopants that form the BAC-Si, BAC-Al, BAC-Ge, and BAC-P were investigated [25, 31, 34]. Following this, the monovalent BACs formed by Bi+ were of definite interest in understanding the NIR emission origin in BDFs. Bi+ was described as having a ground configuration of 6s26p2 and its energy level split by a spin-orbit coupling interaction into the ground state 3P0 and the excited states 1S0, 1D2 and 3P2,1 corresponding to the energy bands around 500 nm, 700 nm, 800 nm and 1000 nm [25, 36]. Recently, energy levels of BAC-Si and BAC-Ge were described. The origin of NIR emission was reported as likely from Bi··· ≡ Si-Si ≡ and Bi··· ≡ Ge-Ge ≡ complexes formed by the interstitial Bi atoms (Bi0) and intrinsic glass defects and ≡ Si-Si ≡ and ≡ Ge-Ge ≡ oxygen vacancies [29, 37]. The energy levels of BAC-Si and BAC-Ge recommended in previous work are presented in **Figure 6** shows the excitation and emission characteristics associated with these particular BACs.

Although BAC-Si and BAC-Ge were described with much information about their kind and spectral characteristics, in-depth knowledge of BAC-Al and BAC-P lags far behind. BAC-Al and BAC-P showed notably different spectral behavior in BDFs than those of BAC-Si and BAC-Ge. One of the essential distinctive features is the Stokes shifts in the emission and excitation produced from these two centers, while for BAC-Si and BAC-Ge, no such Stokes shifts were pronounced [29]. Electron-photon interaction giving increase to the detrimental emission processes joined with these two centers was expected as the reason for

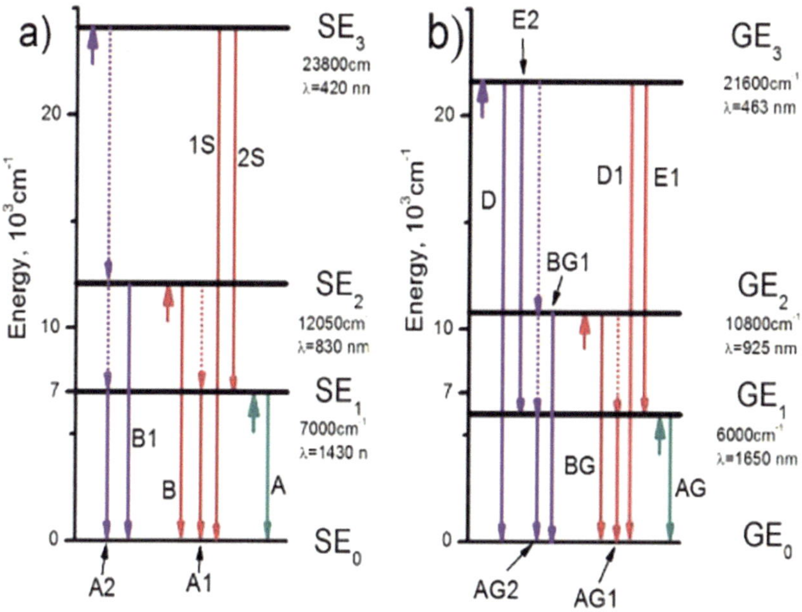

Figure 6.
Reported energy levels of (a) BAC-Si and (b) BAC-Ge [31].

the Stokes shifts. Furthermore, BAC-Al was quite different from BAC-Si and BAC-Ge based on their susceptibility to electron irradiation [38]. All of these specifications made it hard to discover the complete energy diagram of these active centers.

3.2 Luminescence properties of BDFs

The properties of near-infrared (NIR) broadband luminescence of Bi-doped silica fibers are sensitive to the composition of glass in addition to the conditions of fabrication. The partial shielding of the unfilled subshell 4f by filled 5 s and 5p in RE ions prevents significant interaction from the host environment. In comparison to RE elements, Bi has filled internal subshells, and external 6 s and 6p electrons interact significantly with the host, showing host-dependent absorption and emission properties. Co-dopants can also dramatically alter the spectrum of luminescence, as shown in **Figure** 7.

Herein, Bi-doped fibers with aluminosilicate hosts are described as Bi-doped aluminosilicate fibers (BASEs). They have shown a luminescence peak at 1150 nm. In contrast, Bi-doped fibers with phosphorus silicate and Germane silicate core composition hereafter described as Bi-doped phosphorus silicate fiber and Bi-doped German silicate fibers (BGSFs) are known to transfer the emission band to a longer wavelength side of about 1300 and 1450. Recently, by using high GeO2 concentration (50 mol%) within the fiber core, the Bi emission window is extending to a covered wavelength band of 1600–1800 nm. Luminescence in Bi-doped phosphor germane silicate fibers (BPGSFs) is broad compared to phosphorous (P) and other Bi-doped fibers co-doped with germanium (Ge). This makes it possible by changing the fiber core composition using Bi-doped optical fibers to show lasers and amplifiers for cover the full spectrum region about 1150–1800 nm. In **Figure 6**, the pump wavelength bands are referred to by the arrow. Wavelength bands of the pump about 1050, 1230, and 1310 nm for BASFs, BPSFs, and BGSFs,

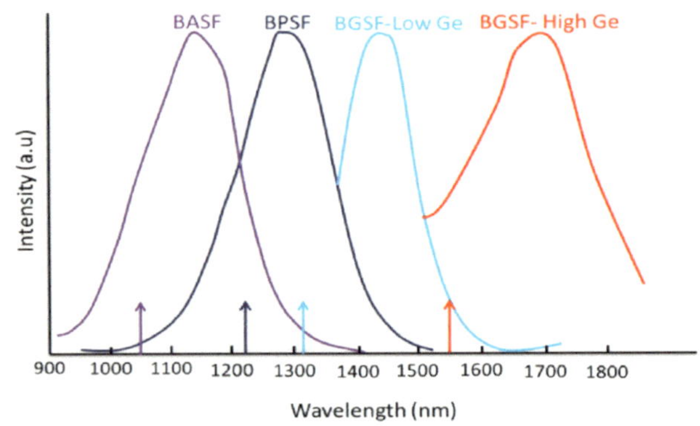

Figure 7.
The Spectrum regions covered via different Bi-doped fibers with various hosts [39].

respectively, usually are utilized. In the case of BGSFs with a high concentration of GeO2, the pump wavelength band is about 1550 nm [39, 40].

3.3 Applications of BDFs

BDFs have many applications in different optical systems because of their wavelength based discriminative transmission losses [41]. A 1651 nm single-frequency laser diode is providing with a Bi-doped fiber power amplifier, and its performance is evaluated experimentally [42]. This amplifier is using to increase the performance of methane detection systems, such as remote stand-off systems or photoacoustic spectroscopy (methane has a molecular absorption line of almost 1651 nm). Two amplifier configurations are shown and output power of larger than 80 mW in both cases. The results obtained provide valuable perspective for compact and straightforward fiber amplifiers in the spectral region of ~1630–1750 nm with a power output of over 100 mW. The experimental set-up schematic diagrams are displayed in **Figure 8(a)** and **(b)**. Two configurations are investigating, both with signal and pump radiation continues to using a WDM coupler to launch pump light at 1550 nm into the BDF and offer the amplified radiation output at 1651 nm. Standard, commercially available WDM coupler that is proposed to operating at 1625 nm and 1550 nm (from the Opto-Link Corporation). Its transmission is calculated to be ~95% at 1550 nm, but only ~65% at 1651 nm. A polarization-independent optical circulator is using in a second configuration (**Figure 8(b)**) broadband as an alternative a WDM coupler. The circulator (including from OptoLink Corporation) has been proposed for a wavelength of 1525–1610 nm. Nevertheless, its 1550 nm (port 1 - port 2) and 1651 nm (port 2 - port 3) transmissions were quite the same, almost 80 percent.

The Er-Doped Fiber Amplifier (EDFA) is currently one of the most crucial elements for various fiber optic systems [43]. Nevertheless, driven by a continually growing demand, network traffic, which has grown exponentially for decades, would result in the overloading of these"capacity crunch" systems, since EDFA is limited to 1530–1610 nm in the spectrum. New technology will need to be pursued, and the development of optical amplifiers for new spectral regions can be a promising solution. Some of the fiber-optic amplifiers are made with materials doped with rare-earth. As a result, full bands for the gain band of Er-doped fibers are still available in shorter (1150–1530 nm) and longer wavelength (1600–1750 nm) regions. A novel fiber amplifier is operated at a spectrum region of 1640–1770 nm pumping via commercial laser diodes at 1550 nm. This amplifier is achieved by use Bi-doped high-Germania silicate fibers fabricating by an MCVD technique. The experimental set-up of a typical Bi-doped fiber amplifier is illustrated schematically in **Figure 9**. The BDFA is built using a scheme with Bi-directional pumping (backward and forward pumping). As a pumping source, commercial laser diodes are using with an ultimate output power of 150 mW each. The active fiber was core-pumped through commercial WDMs based on SMF-28. WDMs Transmission spectra are shown in **Figure 8**. Optical isolators are spliced to the amplifier of the input and output. The first is using to reduce the effect of amplified spontaneous emission (ASE) of a Bi fiber on the signal source. The second prevented possible lasing. Bold points act as splices, where an optical loss is ~1 dB because of the difference between the active fiber mode field diameter and that of conventional fibers (SMF-28).

The proposed dual pumping scheme of 830 and 980 nm aims at broadening and flattening Bi/Er multicomponent fiber (BEDF) spectral performance [44]. The distinct BACs of germanium (BAC-Ge), aluminum (BAC-Al), phosphorus (BAC-P), and silique (BAC-Si) have spectral properties characterized by single pumping, respectively, of

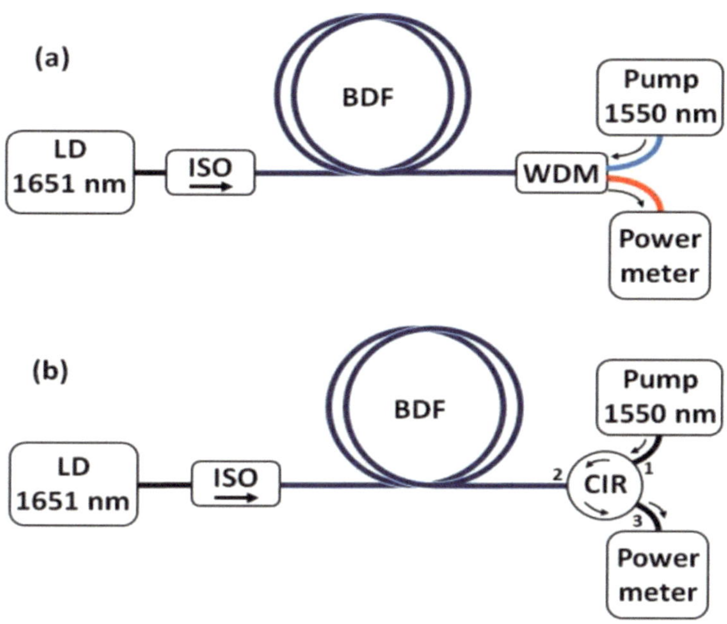

Figure 8.
Two BDF amplifiers configurations: (a) set-up with a WDM coupler using for pump light launch in BDF and processing of amplified signals at 1651 nm; (b) optical circulator set-up. ISO-optical insulator, LD-diode projector.

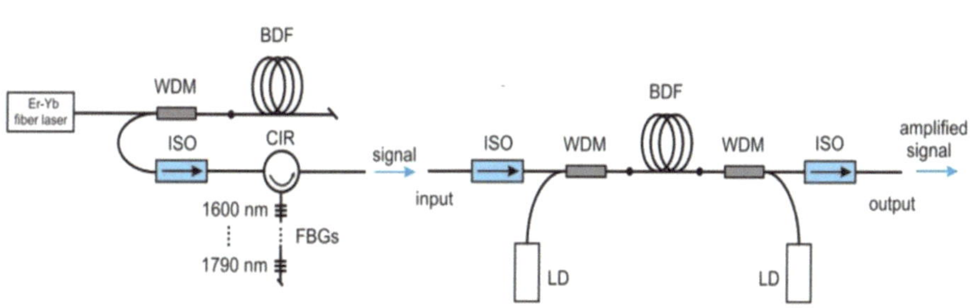

Figure 9.
Experimental BDFA set-up. The figure left side shows a home-built light source producing a wavelength comb of 1615 nm - 1795 nm, which is uniformly spaced with a phase of 15 nm. The figure right side illustrates the amplifier itself. Notably, ISO: An optical isolator, LD: Laser diode operating at 1550 nm, FBG: Fiber Bragg grating and CIR: Optical circulator.

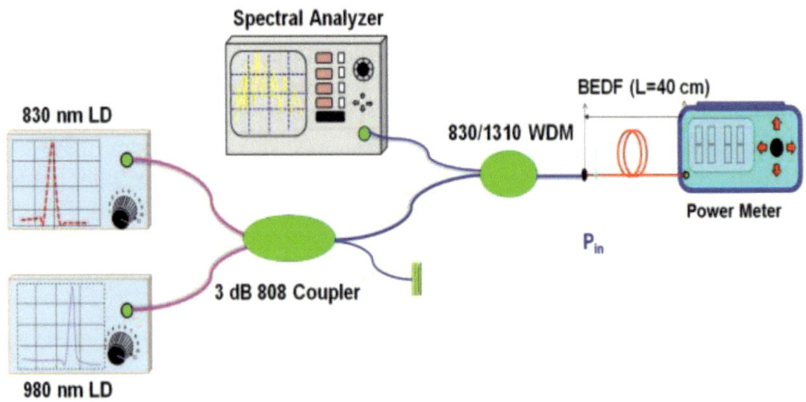

Figure 10.
Dual pumping experimental set-up for backward luminescence measurement.

980 nm and 830 nm. Depending on the BAC-Al (~1100 nm) and BAC-Si (~1430 nm) emission slope efficiencies under the single pumping of 830 and 980 nm, the dual pumping scheme with an optimizing pump power ratio of 25 (980 nm VS 830 nm) is determined to realize flat, ultra-broadband luminescence spectra covered the wavelength range 950-1600 nm. The dual pumping scheme is more illustrated on the on-off gain BEDF performance. The gain spectrum is flattening and broadening over 300 nm (1300-1600 nm) with an average gain coefficient of ~1.5 dBm^{-1}, due to the pump power ratio of ~8 (980 VS 830 nm). The spectral covering is about 1.5 and 3 times more expansive compared to single pumping of 830 and 980 nm pumping, respectively. Due to the optical characteristic of the energy level diagrams of 830 and 980 nm, the advantage of dual pumping is clarified. The proposed dual 830 and 980 nm pump scheme with the BEDF multicomponent shows great promise potential in a range of broadband optic applications, such as NIR-band tunable laser, a standard ASE source, and broadband amplifier. The luminescence measurement set-up is shown in **Figure 10**. The 980 and 830 nm pigtailed laser diodes (LD) were launched into the input end of 3 dB 808 couplers, and the 810/1310 WDM, the BEDF, was spliced with the output end of 1310 beam with ~1 dB splice loss. The backward optical spectral analyzer (OSA) was used to record the emission signal to remove the effect of residual pump power. To monitor the unabsorbed pump power, a digital power meter was placed at the end of BEDF, and A short length (~40 cm) of BEDF was tested.

4. Bismuth doped fiber amplifier (BDFA)

Doped fiber amplifiers (DFAs) are optical amplifiers that use a doped optical fiber as a gain medium to amplify an optical signal. They are related to fiber lasers. The pump laser and the signal to be amplified are multiplexed into the doped fiber, and the signal is amplified through interaction with the doping ions. As an example is the (BDFA), where the core of a silica fiber is doped with Bi ions and can be efficiently pumped with a laser. 1120 nm diode-pumped Bi-doped fiber amplifier is fabricated by N. K. Thipparapu et al. [45]. Bi-doped aluminosilicate fiber is fabricated using an MCVD solution doping and is distinguished by its gain and unsaturated loss. The amplifier performance was compared with the traditional pumping wavelength region of 1047 nm for a novel pumping wavelength of 1120 nm. Unsaturable losses in 1047 and 1120 nm pump wavelengths were 65% and 35%, respectively. At 1180 nm, the maximum gain of about 8 dB was observed for 100 m fibers with pumping at 1120 nm. The 1120 nm pump produced an enhancement of gain of 70% compared with the 1047 nm pump. An additional 3.5 dB gain was achieved at 1047 and 1120 nm simultaneous pumping. **Figure 11** shows the schematic experimental set-up used for the measurement of gain in Bi-doped fiber. The set-up consists of an 1180 nm LD as input signal source and fiber pigtailed 1120 nm LD and/or 1047 nm Nd-YLF laser as the pump source. To abstain from the wavelength division multiplexers (WDMs) and back reflections to combining pump and signal sources, Isolators (ISO) was used. The spectrum of the output and input signal is taking through an OSA. The input signal was obtained little before the fiber under trial. In contrast, the output signal is determined by consideration of the WDM loss using to separating the pump from the signal.

O-band has recently been widely used for low-cost data transmission. The feature of O-band is that the transmitter wavelength(s) are laying close to the zero dispersive fiber wavelength (λ_0), and no compensation is therefore needed for either optical or electronic chromatic dispersion. O-band transponder total bit rate is enhanced to 425 Gb/s, for instance, using 8 LAN WDM 26.6 Gbaud/s PAM-4 modulated channels [46]. Using WDM and complex modulation formats minimize the power of both the receiver sensitivity and receiver per channel such that optical amplification is appropriate. O-band signal boosting can be achieved by employing Semiconductor Optical Amplifiers (SOA). However, they create considerable distortions due to the modulation of self- and cross- gain [47]. Although most of dispersion broadened channels are amplified [48], SOAs are not approbated for intensity modulation formats transmission like PAM-4 operating near λ0 with relatively small channel count. Praseodymium doped fiber amplifiers (PDFA) with a bandwidth of 1280–1320 nm [49] were shown, but non-silica host glass was required, which makes PDFA complicated and costly. O-band amplification has been extensively studied in Bismuth-doped silica fibers [50]. A 150 m long BDFA with bandwidths of 1320-1360 nm has been reported using a complex dual-wavelength pumping system with only 6 × 10 Gb/s OOK channels [51]. Simple silica-based BDFA with 80 nm 6-dB gain-bandwidth flexibly centered within 1305-1325 nm, and parameters comparable to EDFAs is developed by V. Mikhailov et al. [52]. The amplifier can extend 400GBASE-LR8 transmission (8 × 26.6 Gbaud/s PAM-4 channels) beyond 50 km of G.652 fiber. The active fiber core consists of phosphosilicate glass doped with Bi (>0.01 mol%) produced using the MCVD process. The preform was cladding by a Heraeus F300 tube while all the core components, including P, Si, and Bi, were deposited from the gas phase. The index difference between the fiber core and the cladding was approx. 6 × 10–3, and the cut-off wavelength was near 1.1 μm. As the core diameter of 7 μm of fiber offered reasonable slice capability with silica-based fibers, the regular automatic splicer was

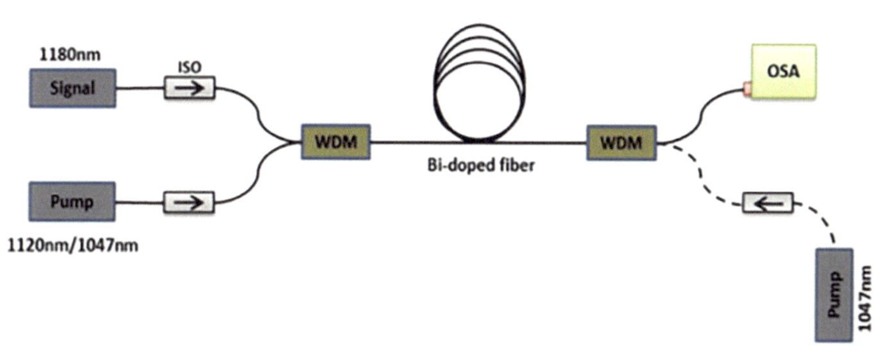

Figure 11.
Schematic of the experimental set-up to measure gain in Bi-doped fiber (the 1047 nm pump with the dashed line is used for Bi-directional pumping; otherwise, that port of the WDM is used to monitor the excess pump).

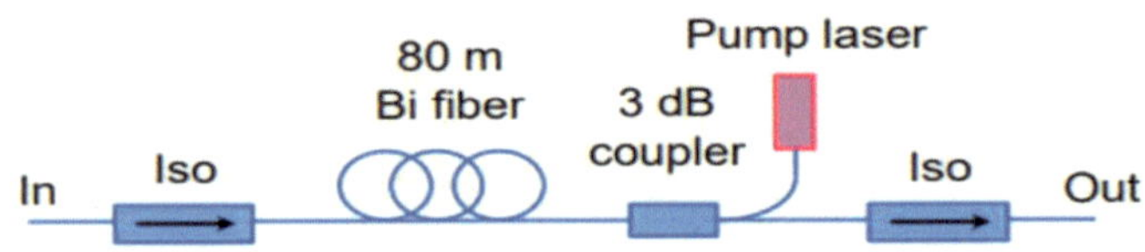

Figure 12.
Bi-fiber characterization set-up.

used to splice G.652and Bi- fibers. In order to investigate emission properties, 80 meters of Bi-fiber were subsequently pumped with a 3 dB wideband coupler with lasers of 1155, 1175, 1195, 1215, and 1235 nm (**Figure 12**).

Author details

Rifat M. Dakhil Alsingery* and Ahmed Mudhafer
Chemical and Petrochemical Techniques Engineering Department, Southern Technical University, Iraq

*Address all correspondence to: rifat.dakhil@stu.edu.iq

References

[1] K. Kao and G. A. Hockham, "Dielectric fiber surface waveguides for optical frequencies," in *Proceedings of the Institution of Electrical Engineers*, vol. 113, no.7.IET, 1966, pp. 1151-1158

[2] F. Kapron, D. Keck, and R. Maurer, "Radiation losses in glass optical waveguides," *Applied Physics Letters*, vol. 17, no. 10, pp. 423-425, 1970.

[3] T. Miya, Y. Terunuma, T. Hosaka, and T. Miyashita, \Ultimate low-loss single mode fiber at 1.55μm," Electronics Letters, vol. 15, no. 4, pp. 106{108, 1979

[4] OFS. All Wave One Fiber-Zero Water Peak. [Online]. Available: http://http:// fiber-optic-catalog.ofsoptics.com/Asset/ AllWave-One-Fiber-160-web.pdf

[5] R. J. Mears, L. Reekie, I. Jauncey, and D. N. Payne, Low-noise erbium-doped fiber amplifier operating at 1.54μm," Electronics Letters, vol. 23, no. 19, pp. 1026-1028, 1987.

[6] D. Richardson, J. Fini, and L. Nelson, Space-division multiplexing in optical fibers," Nature Photonics, vol. 7, no. 5, pp. 354-362, 2013.

[7] Internet use reaches 5 billion worldwide. [Online]. Available: http://www.futuretimeline. net/21stcentury/2020. htm#internet-2020

[8] L. Technologies. All Wave Single Mode Optical Fiber. [Online]. Available:https://www.usbid.com/ datasheets/usbid/2000/2000-q3/5822-3. pdf

[9] Thipparapu, Naresh Kumar. Development of bi-doped fiber amplifiers and lasers & broadband Er-doped multi-element fiber amplifiers. Diss. University of Southampton, 2018.

[10] L. Dong and B. Samson, Fiber Lasers: Basics, Technology, and Applications. CRC Press, 2016.

[11] P. C. Becker, A. A. Olsson, and J. R. Simpson, Erbium-doped fiber amplifiers: fundamentals and technology. Academic Press, 1999.

[12] A. B. G. Blase, "Investigations on Bi^{3+}-activated phosphors" *J. Chem. Phys.*, vol. 48, pp. 217-222, 1968.

[13] R. B. Laurer, "Photoluminescence in $Bi1_2SiO_{20}$ and $Bi1_2GeO_{20}$," *Appl. Phys. Lett.*, vol. 17, pp. 178-179, 1970

[14] M. J. Weber and R. R. Monchamp, "Luminescence of $Bi_4Ge_3O_{12}$: spectral and decay properties" *J. Appl. Phys*, vol. 44, pp. 5495-5499, 1973

[15] F. Kellendonk, T. Belt and G. Blasse, "On the luminescence of bismuth, cerium, and chromium and ytterbium aluminium borate" *J. Chem. Phys.*, vol. 76, pp. 1194-1201, 1982.

[16] R. Retoux, F. Studer, C. Michel, B. Raveau, A. Fontaine and E. Dartyge, "Valence state for bismuth in the superconducting bismuth cuprates" *Phys Rev. B*, vol. 41, pp. 193-199, 1990.

[17] M. A. Hamstra, H. F. Folkerts, and G. Blasse, "Red bismuth emission in alkaine-earthmetal sulphates" *J. Mater. Chem.*, vol. 4, pp. 1349-1350, 1994

[18] Y. Fujimoto and M. Nakatsuka, "Infrared luminescence from bismuth-doped silica glass" *Jpn. J. Appl. Phys.*, vol. 40, pp. L279-L281, 2001.

[19] M. Y. Peng, J. R. Qui, D. P. Chen, X. G. Meng, I. Y. Yang, X. W. Jiang, et al., "Bismuth- and aluminum-codoped germanium oxide glasses for super-broadband optical amplification" *Opt. Lett.*, vol. 29, pp. 1998-2000, 2004

[20] Y. Fujimoto and M. Nakatsuka, "Optical amplification in bismuth-doped silica glass, " *Appl. Phys. Lett.*, vol. 82, pp. 3325-3326, 2003.

[21] E. M. Dianov, V. V. Dvoyrin, V. M. Mashinsky, M. V. Yashkov, and A. N. Guryanov, "CW bismuth fiber laser" *Quantum Electron.*, vol. 35, pp. 1083-1084, 2005.

[22] V. V. Dvoyrin, V. M. Mashinsky, E. M. Dianov, A. A. Umnikov, M. V. Yashkov and A. N. Guryanov, "Absorption, fluorescence and optical amplification in MCVD bismuth-doped silica glass optical fibres" in *Proc. Of the ECOC*, vol. 4, pp. 949-950, 2005.

[23] Y. Fujimoto and M. Nakatsuka, "27Al NMR structural study on aluminum coordination state in bismuth-doped silica glass" *J. Non-Cryst. Solids*, vol. 352, pp. 2254-2258, 2006.

[24] T. Suzuki and Y. Ohishi, "Ultrabroadband near-infrared emission from Bi-doped Li2O–Al2O3–SiO2 glass" Appl. Phys. Lett., vol. 88, p. 191912, 2006.29

[25] X. Meng, J. Qiu, M. Peng, D. Che, Q. Zhao, X. Jiang, et al., "Near infrared broadband emission of bismuth-doped aluminophosphate glass" *Opt. Express*, vol. 13, pp.1628-1634, 2005.

[26] M. Peng, X. Meng, J. Qiu, Q.Zhao and C. Zhu, "GeO2: Bi, M (M = Ga, B) glasses with super-wide super-wide infrared luminescence" *Chem. Phys. Lett.*, vol. 403, pp. 410-414, 2005.

[27] S. V. Firstov, I. A. Bufetov, V. F. Khopin, A. A. Umnikov, A. N. Guryanov, and E. M. Dianov, "Time-resolved spectroscopy and optical gain of silica-based fibers co-doped with Bi, Al and/or Ge, P, and Ti" *Laser Phys.*, vol. 19, pp. 894-901, 2009.

[28] Y. Fujimoto, "Local structure of the infrared bismuth luminescent center in bismuthdoped silica glass" *J. Am. Ceram. Soc.*, vol. 93, pp. 581-589, 2010.

[29] I. A. Bufetov, M. A. Melkumov, S. V. Firstov, K. E. Riumkin, A. V. Shubin, V. F. Khopin, et al., "Bi-Doped Optical Fibers and Fiber Lasers" *IEEE J. Sel. Top. Quant.* Vol. 20, 0903815, 2014.

[30] I. A. Bufetov, S. V. Firstov, A. V. Shubin, S. L. Semenov, V. V. Vel" Miskin, A. E. Levchenko, et al., "Optical gain and laser generation in bismuth-doped silica fibers free of other dopants" *Opt. Lett.*, vol. 36, pp. 166-168 2011.

[31] S. V. Firstov, V. F. Khopin, I. A. Bufetov, E. G. Firstova, A. N. Guryanov, and E. M. Dianov, "Combined excitation-emission spectroscopy of bismuth active centers in optical fibers" *Opt. Express*, vol. 19, pp. 19551-19561 2011.

[32] E. M. Dianov, "Bismuth-doped optical fibers: a challenging active medium for near-IR lasers and optical amplifiers" *Light Sci. Appl.*, vol. 1, pp. 1-7, 2012.

[33] E. M. Dianov, "Bismuth-doped optical fibers: a new active medium for NIR lasers and amplifiers" in *Proc of the SPIE*, vol. 8601, 2013.

[34] T. M. Hau, X. Yu, D. Zhou, Z. Song, Z. Yang, R. Wang, *et al.*, "Super broadband near-infrared emission and energy transfer in Bi–Er co-doped lanthanum aluminosilicate glasses" *Opt. Mater.*, vol. 35, pp. 487-490, 2013.

[35] A. N. Romanov, Z. T. Fattakhova, A. A. Veber, O. V. Usovich, E. V. Haula, V. N. Korchak, *et al.*," On the origin of near-IR luminescence in Bi-doped materials (II). Subvalent monocation Bi+ and cluster Bi53+ luminescence in AlCl3/ZnCl2/BiCl3 chloride glass," Opt. Express, vol 20, pp. 7212, 2012. 30

[36] L. Zhang, G. Dong, J. Wub, M. Peng, and J. Qiua, "Excitation wavelength-dependent near-infrared luminescence

from Bi-doped silica glass" *J. Alloy Compd.*, vol. 531, pp. 10-13, 2012.

[37] V. O. Sokolov, V. G. Plotnichenko, and E. M. Dianov, "The origin of near-IR Luminescence in bismuth-doped silica and Germania glasses-free of other dopants: firstprinciple study" *Opt. Mater. Express*, vol. 3, pp. 1059-1074, 2013.

[38] A. V.Kir'yanov, V. V. Dvoyrin, V. M. Mashinsky, N. N. Il' ichev, N. S. Kozlova, and E. M. Dianov, "Influence of electron irradiation on optical properties of Bismuth doped silica fibers" *Opt. Express*, vol. 19, pp. 6599-6608, 2011.

[39] I. Bufetov and E. Dianov, Bi-doped fiber lasers,“ *Laser Physics Letters*, vol. 6, no. 7, p. 487, 2009.

[40] S. Firstov, S. Alyshev, M. Melkumov, K. Riumkin, A. Shubin, and E. Dianov, Bismuth-doped optical fibers and fiber lasers for a spectral region of 1600-1800nm,“ *Optics Letters*, vol. 39, no. 24, pp. 6927-6930, 2014.

[41] D. Jain, N. K. Thipparapu, and J. K. Sahu, "Bismuth doped fiber for filtering applications," in 2019 Conference on Lasers and Electro-Optics Europe and European Quantum Electronics Conference, OSA Technical Digest (Optical Society of America, 2019), paper cj_p_39.

[42] Michal Nikodem, *Aleksandr M Khegai and Sergei V Firstov*' "Single-frequency bismuth-doped fiber power amplifier at 1651 nm", 2019 *Laser Phys. Lett.* **16** 115102

[43] Firstov, S. V., Alyshev, S. V., Riumkin, K. E., Khopin, V. F., Guryanov, A. N., Melkumov, M. A., & Dianov, E. M. (2016). A 23-dB bismuth-doped optical fiber amplifier for a 1700-nm band. *Scientific reports*, 6, 28939.

[44] Zhao, Q., Luo, Y., Wang, W., Canning, J., & Peng, G. D. (2017). Enhanced broadband near-IR luminescence and gain spectra of bismuth/erbium co-doped fiber by 830 and 980 nm dual pumping. *AIP Advances*, *7*(4), 045012.

[45] Thipparapu, N. K., Jain, S., Umnikov, A. A., Barua, P., & Sahu, J. K. (2015). 1120 nm diode-pumped Bi-doped fiber amplifier. *Optics Letters*, *40*(10), 2441-2444.

[46] IEEE 802.3bs-2017 - IEEE Standard for Ethernet Amendment 10: Media Access Control Parameters, Physical Layers, and Management Parameters for 200 Gb/s and 400 Gb/s Operation, IEEE Standard, 2017.

[47] A.A.M Saleh et al., Effects of semiconductor optical amplifier nonlinearity on the performance of high-speed intensity modulation lightwave systems, IEEE Trans. Com., vol. 38, 1990.

[48] J. Renaudier, 100nm ultra-wideband optical fiber transmission systems using semiconductor optical amplifiers, Proc. ECOC, 2018, Mo4G.5.

[49] M. A. Melkumov, et al., Bismuth-doped fiber lasers and amplifiers: Review and prospects, Proc. Int. Conf. Laser Optics, 2016, S1-19.

[50] S. Barthomeuf, et al., High Optical Budget 25Gbit/s PON with PAM4 and Optically Amplified O-Band Downstream Transmission, Proc. ECOC, 2018, Mo4B.2.

[51] N. Taengnoi, et al., Amplified O-band WDM Transmission using a Bi-doped Fibre Amplifier, Proc. ECOC, 2018, Mo3E.2.

[52] Mikhailov, V., Melkumov, M. A., Inniss, D., Khegai, A. M., Riumkin, K. E., Firstov, S. V., ... & Puc, G. S. (2019, March). Simple broadband bismuth-doped fiber amplifier (BDFA) to extend O-band transmission reach and capacity. In *Optical Fiber Communication Conference* (pp. M1J-4). Optical Society of America.

Chapter 3

BAC Photobleaching in Bismuth-Doped and Bismuth/Erbium Co-Doped Optical Fibers

Bowen Zhang, Mingjie Ding, Shuen Wei, Binbin Yan, Gang-Ding Peng, Yanhua Luo and Jianxiang Wen

Abstract

Bismuth-doped optical fiber (BDF) and bismuth/erbium co-doped optical fiber (BEDF) have attracted much attention due to their ultra-broadband luminescence in the near-infrared (NIR) region. The photobleaching effect on bismuth active centers (BACs) related to the NIR luminescence has been systematically investigated and summarized, in terms of irradiation intensity, irradiation wavelength, and temperature. All these findings not only give the deep insights into the fundamental structure of BACs but also provide an effective way to control the BACs. They play an important role for the development of BDF- and BEDF-based devices with high performance and stability under laser exposure in future.

Keywords: bismuth-doped optical fiber (BDF), bismuth/erbium co-doped optical fiber (BEDF), bismuth active center (BAC), laser irradiation, photobleaching, irradiation intensity, irradiation wavelength, temperature

1. Introduction

Since Fujimoto et al. in 1999 first demonstrated that bismuth-doped silica glass could generate broadband luminescence covering the near-infrared (NIR) region [1], the bismuth-doped materials have attracted considerable attention due to their ultra-wide luminescence [2–8]. Especially, the bismuth-doped and bismuth/erbium co-doped optical fibers (BDFs and BEDFs) have been developed for tunable fiber laser, amplifier, and ultra-broadband light source operating in the range from 1000 to 1800 nm [9–15].

Although great endeavor has been made to improve the performance of BDFs and BEDFs, challenges still exist, and they have become obstacles to many practical applications. One of the key challenges is that the fundamental structure of bismuth active centers (BACs) and the nature of their NIR luminescence remain unclear. Based on the previous researches, it was generally accepted that the formation of BACs greatly depends on the material compositions [12, 15]. With different doping elements such as aluminum, phosphorus, silicon and germanium in the glass environment, there are four types of BACs in the BDFs, namely BAC associated with aluminum (BAC-Al), BAC associated with phosphorus (BAC-P), BAC associated with silicon (BAC-Si), and BAC associated with germanium (BAC-Ge), respectively [16–19]. Multiple absorption peaks of these BACs have been observed in BDFs and BEDFs, for example, BAC-Al (510, 700, and 1050 nm), BAC-P (460, 750, and 1300 nm), BAC-Si (420, 830, and 1400 nm), and BAC-Ge (463, 925, and 1600 nm) [17, 18]. The typical NIR luminescence bands of these BACs locate at ~1100 nm (BAC-Al), ~1300 nm (BAC-P), ~1400 nm (BAC-Si), and ~ 1700 nm (BAC-Ge), respectively [15, 17].

To better reveal the nature of the NIR luminescence in BDFs/BEDFs and the configuration of BACs, various post treatments, such as ionizing radiation, laser irradiation, and thermal treatment, have already been applied [20–25]. Thereinto, as a result of laser irradiation, the photobleaching of BDFs/BEDFs leading to the gradual decay of the luminescence of BACs is quite obvious. In this chapter, the photobleaching effect on BACs observed in BDFs and BEDFs has been reviewed. More specially, this effect is demonstrated and analyzed in detail from the angle of BAC type, irradiation intensity, irradiation wavelength, and temperature. In addition, the photobleaching mechanism for each BAC is also discussed. The investigation of this photobleaching process gives not only the deep insights into the structure of BACs but also more information of photostability of BDFs/BEDFs. With further understanding of the BACs, it helps to develop an effective way to control the BACs and obtain better and more stable optical performance of BDF and BEDF for practical applications.

2. Phenomenon of photobleaching

The photobleaching effect observed in some luminescent materials is featured by the gradual decay of luminescence after laser irradiation. This effect can be referred as a process of laser irradiation-induced luminous centers destruction and/or converting into the nonluminous center. This process is called the photobleaching effect [26].

The photobleaching effect has been found in many materials. For example, it was reported that the green fluorescent protein could be bleached under laser irradiation [27]. The Sm^{2+} emission in epitaxial CaF_2 film could be bleached partially under 633-nm irradiation [28]. The similar photoinduced reduction of luminescence has also been observed in Nd^{3+}:$LiYF_4$ and YVO_4:Bi^{3+}/Eu^{3+} nanoparticles [29, 30]. In addition, a large number of dyes present the photobleaching characteristics [31–34]. In general, the photobleaching effect is caused by breaking of covalent bonds or nonspecific reactions between the luminous center and surrounding molecules. Especially, some photobleaching-based techniques such as fluorescence loss in photobleaching and fluorescence recovery after photobleaching have been developed and used for molecular marker, in vivo cell tracking, and investigation of molecule diffusion in biology [27, 35–37].

The photobleaching effect also exists in the glass materials. Exposure to 977 nm light irradiation led to the absorption decrease of Yb^{3+} in ytterbium-doped silica fiber [38]. In addition, the darkened Yb-doped fiber could be photobleached when irradiated by 355 and 633 nm laser [39, 40]. The similar photobleaching effect has also been observed in the thulium-doped fibers [41].

As for the bismuth-doped materials, the BAC-related NIR luminescence was reduced under laser irradiation in bismuth-doped silica-based glasses, $TlCdCl_3$ crystal and $Sr_2B_5O_9Cl$:Bi crystal [42–44]. In the case of BDFs/BEDFs, a number of studies on the photobleaching of various BACs have been reported [19, 21, 45–54]. Herein, the photobleaching effect on BAC-Si, BAC-Ge, BAC-Al, and BAC-P in BDFs and BEDFs is demonstrated in relation to irradiation intensity, irradiation wavelength, and temperature, respectively. The underlying mechanisms of these photobleaching effects in BDFs and BEDFs are also discussed below.

3. Photobleaching of BAC-Si

The photobleaching of BAC-Si in BEDF under 830-nm laser irradiation has been reported in [46]. By the 830 nm laser irradiation with an intensity of 0.12 MW/cm^2, the luminescence spectra of 50 cm BEDF at different irradiation time were shown in **Figure 1a**. It is evident that the NIR luminescence of BAC-Si peaking at ~1420 nm decreases continuously with the longer exposure time. In addition, the absorption of BAC-Si peaking at ~816 nm demonstrates the similar trend, gradually decaying when exposed to the 830 nm laser as shown in **Figure 1b**. The decrease of both the NIR luminescence and absorption clearly indicates the degradation of BAC-Si. Interestingly, after the irradiation, both the absorption and luminescence of BAC-Si gradually recover to the initial value in 48 hours at room temperature (RT). This recovery behavior implies that the photobleaching of BAC-Si under 830-nm irradiation in BEDF is reversible in the mild condition.

To characterize the photobleaching effect on BAC-Si, the stretched exponential function (SEF) has been used, which is expressed as [55]:

$$I(t) = I_1 + (I_0 - I_1)e^{-\left(\frac{t}{\tau}\right)^{\beta}} \quad (1)$$

where I_1 and I_0 are the luminescence intensities at bleaching saturated time and initial time, τ represents the bleaching time and β stands for the stretched parameter. In addition, the bleaching ratio r_B, defined by bleached part of the luminescence (I_0–I_1) divided by the initial luminescence before irradiation (I_0), is used to quantify the photobleaching degree [47]. **Figure 2** shows the typical luminescence variation at 1420 nm irradiated by 0.12 MW/cm^2 830-nm laser, which is well fitted by SEF in Eq. (1).

3.1 Irradiation intensity dependence

It has been found that the irradiation intensity (power) has a significant effect on the photobleaching of BACs. Higher irradiation intensity provides more photons, which may lead to stronger photobleaching effect. The irradiation intensity dependence of photobleaching of BAC-Si in BEDF was investigated under 830-nm laser irradiation, and the variation of irradiation intensity was achieved by changing the incident power ranged from 0.39 to 35 mW [47]. **Figure 3a** demonstrates the time

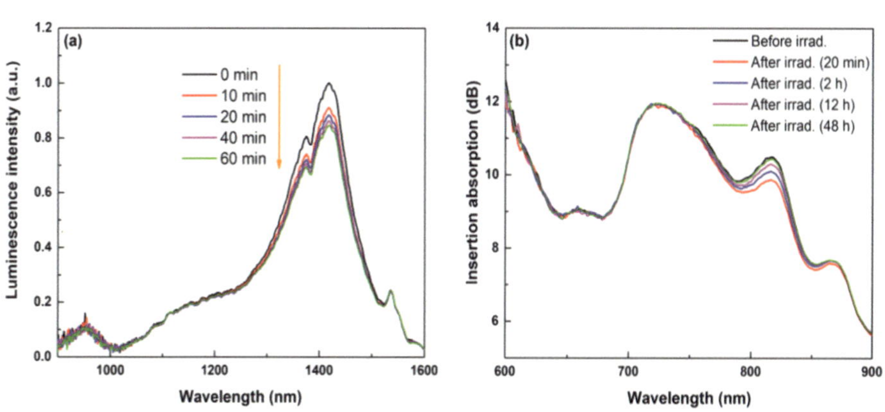

Figure 1.
(a) Luminescence spectra of the BEDF under 830-nm irradiation in 60 minutes of exposure time; (b) BEDF insertion absorption spectra before and after irradiation (20 minutes to 48 hours) [46].

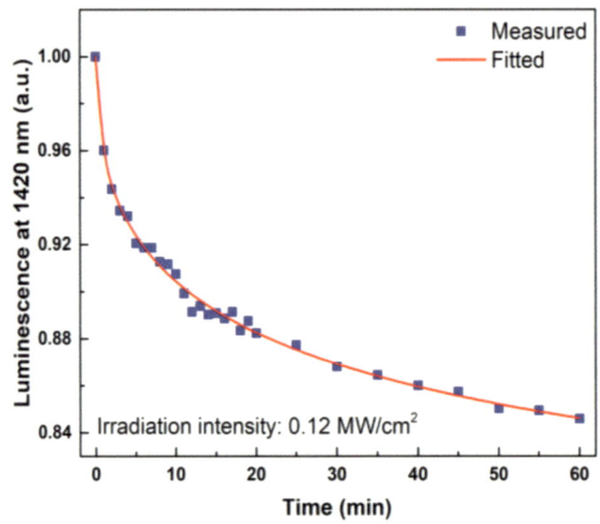

Figure 2.
Variation of luminescence intensity at 1420 nm as a function of irradiation time [46].

evolution of BAC-Si luminescence at 1420 nm under different irradiation power. It is obvious that the luminescence of BAC-Si decays more severely as the irradiation power increases. Furthermore, the luminescence at 1420 nm, time constant $1/\tau$, and bleaching ratio r_B vs. irradiation power is plotted as **Figure 3b**. It is worth noting that both the bleaching ratio and bleaching rate tend to be saturated as the irradiation power increases. Such trend is similar with the variation trend of luminescence at 1420 nm. It hints that the excitation of BAC-Si may participate in the photobleaching process.

3.2 Irradiation wavelength dependence

The photobleaching of BAC-Si was observed in BEDF under 710 and 1380 nm irradiation. As shown in **Figure 4**, the luminescence of BAC-Si is significantly bleached under irradiation of 710 nm, but almost no change under 980-nm irradiation. The inset of **Figure 4** indicates that the BAC-Si luminescence can be slightly bleached under 1380-nm irradiation, which is much weaker than that 710 nm irradiation. In addition, the luminescence of BAC-Si in BDF can be bleached under 532- and 407-nm irradiation [45]. The photobleaching effect on BAC-Si with different irradiation wavelengths is further summarized and listed in **Table 1**. Seen from **Table 1**, except for 980 nm, the photobleaching of BAC-Si can be obtained under all the other applied irradiation wavelengths even the irradiation intensity for some wavelengths is quite small. Generally, it is believed that the shorter wavelength provides larger photon energy. With the increasing photon energy, a greater number of BACs are degraded, resulting in the stronger photobleaching effect. However, even though light at 980 nm has more photon energy than that at 1380 nm, no obvious photobleaching effect is observed under 980-nm irradiation. Considering the fact that 980 nm is unable to excite BAC-Si to the upper energy level, therefore, it is suggested that the photobleaching effect on BAC-Si can only happen when the irradiation wavelength is capable of pumping BAC-Si to excited state. The detailed mechanism will be discussed in Section 3.4.

3.3 Temperature dependence

The temperature dependence of photobleaching effect on BAC-Si in BEDF was studied in the range from 77 to 673 K [47]. The bleaching ratio of BAC-Si under

0.36 MW/cm^2 830-nm irradiation as a function of temperature is plotted as **Figure 5**. As the temperature rises from 77 to 673 K, the bleaching ratio of BAC-Si increases from 34 to 66% and then tends to be saturated. This result indicates that the heating of BEDF

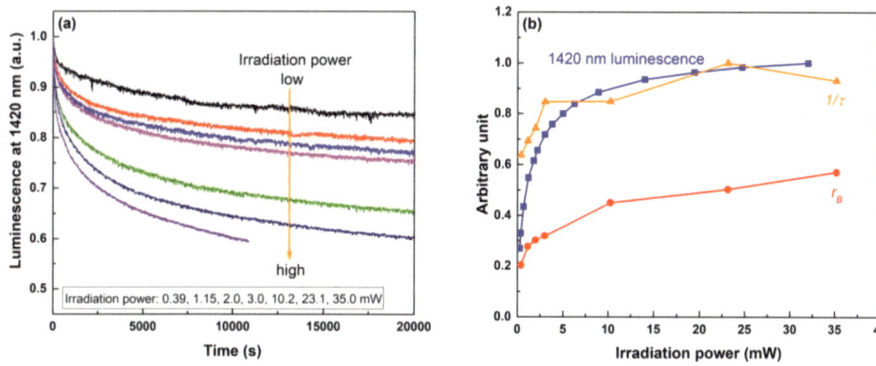

Figure 3.
(a) Evolution of BAC-Si luminescence at 1420 nm under different irradiation power; (b) luminescence, time constant, and bleaching ratio vs. irradiation power [47].

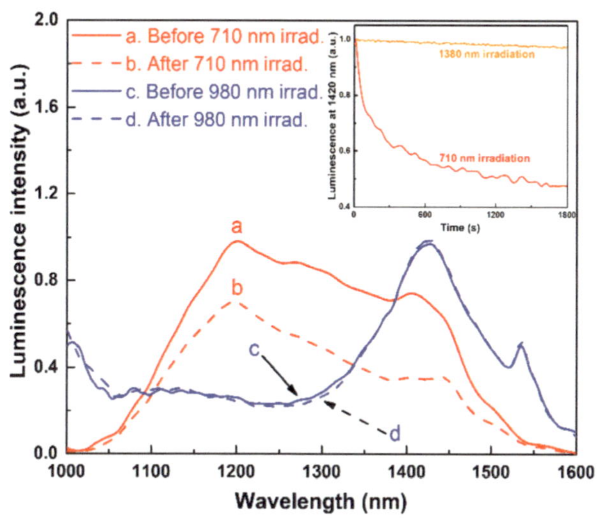

Figure 4.
Luminescence spectra excited by 0.2 mW 830-nm laser before and after 710- and 980-nm irradiation. The inset presents the variation of luminescence at 1420 nm with irradiation wavelengths of 710 and 1380 nm [47].

Sample	Irradiation λ (nm)	Intensity (MW/cm^2)	r_B	Ref.
BDF	407	1	0.85	[45]
BDF	532	1.5	0.8	[45]
BEDF	710	0.36	0.55	[47]
BEDF	830	0.36	0.57	[47]
BEDF	980	0.36	—	[47]
BEDF	1380	0.005	0.06	[47]

Table 1.
Summary of irradiation wavelength influence upon photobleaching of BAC-Si.

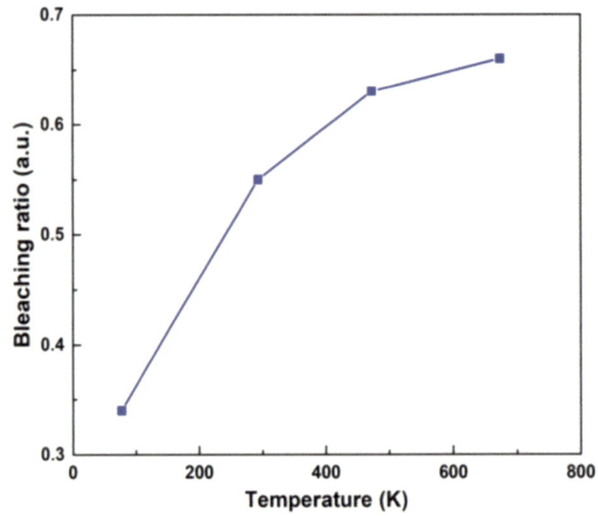

Figure 5.
Temperature dependence of bleaching ratio of BAC-Si under 0.36 MW/cm² 830-nm irradiation [47].

to the high temperature can accelerate the electron movement, leading to the stronger photobleaching of BAC-Si.

3.4 Photobleaching mechanism of BAC-Si

Based on the fact that the excitation of BAC-Si is an essential condition for the photobleaching [47], the photobleaching process is deduced to be expressed as:

$$Bi^{+x}_{BAC-Si} \xrightarrow{h\nu} {}^{*}Bi^{+x}_{BAC-Si} \xrightarrow{kT} Bi^{+(x+1)}_{BAC-Si} + e^{-}, \quad (2)$$

where Bi^{+x}_{BAC-Si} stands for the Bi ion in a BAC-Si with a valence state of +x, '*' symbolizes the excited state of the active center, and $h\nu$ and kT are the photon energy and thermal energy, respectively. As shown by Eq. (2), the photobleaching mechanism of BAC-Si can be described as: first, the Bi ions absorb the 830 nm photons and are pumped to the upper excited state corresponding to 816 nm; second, with the help of thermal energy, some Bi ions release the electrons which are seized by the surrounding material defects (e.g., self-trapped holes, STH), arousing the decay of luminescence and absorption. As seen from **Table 1**, the BAC-Si could also be bleached under 1380-nm irradiation with a slight bleaching ratio of 0.06. The Bi ions absorb 1380 nm photons and then are excited to the lower excited state, and then the electrons can transfer from a small part of Bi ions at the lower excited state to the nearby defects. Compared with Bi ions at the upper excited state, this electron movement pathway at the lower excited state is more difficult.

4. Photobleaching of BAC-Ge

Photobleaching of BAC-Ge has been found in bismuth-doped optical germanosilicate fibers [45]. After 30 minutes' irradiation of 532-nm laser with an intensity of 1.2 MW/cm^2, both the absorption and the luminescence related to BAC-Ge significantly change as shown by the absorption and luminescence spectra of BDF before and after the irradiation. The dramatical decrease of absorption at 925 and 1650 nm and luminescence at 1700 nm clearly indicate the destruction (photobleaching) of BAC-Ge.

The photobleaching of BAC-Ge can also be fitted well by SEF in Eq. (1). The variation of luminescence at 1700 nm excited by 1550 nm under 532-nm irradiation has been demonstrated in [19, 45]. The decay curve of BAC-Ge luminescence at 1700 nm shows the exponential trend. Moreover, there is no recovery behavior after the photobleaching at RT. However, by thermal annealing of the bleached BDF, BAC-Ge could be recovered [19, 45, 53]. After 532-nm laser irradiation, the luminescence of BAC-Ge almost disappears. Annealing the bleached BDF at 600°C, the luminescence of BAC-Ge at ~1700 nm significantly increases, even more than the pristine value, as shown in **Figure 6**. The evident increase of the luminescence of BACs by thermal treatment has been observed in unirradiated BDFs and BEDFs [22, 24, 25, 56–59]. All these results indicate that the thermal treatment can not only lead to the recovery of irradiated fibers but also activate the new BACs.

4.1 Irradiation intensity dependence

The irradiation intensity dependence of photobleaching of BAC-Ge has been studied under 532-nm irradiation with intensity ranged from 0.5 to 2 MW/cm^2 [48]. The bleaching rate $1/\tau$ calculated by the SEF as a function of irradiation power under 532-nm irradiation is plotted as **Figure 7**. Seen from **Figure 7**, the relationship between the bleaching rate and irradiation power in log-log scale is almost linear with a slope of ~1.7. The fitting slope is close to 2, which suggests that the photobleaching of BAC-Ge under 532-nm irradiation is likely to be a two-photon process. In addition, the bleaching rate increases with the irradiation intensity, indicating that more photons obtained by higher irradiation intensity results in faster photobleaching of BAC-Ge.

4.2 Irradiation wavelength dependence

The irradiation wavelength dependence of photobleaching of BAC-Ge differs from that of BAC-Si. As mentioned above, the photobleaching of BAC-Si only happens when the irradiation photon is able to excite BAC-Si. The photobleaching

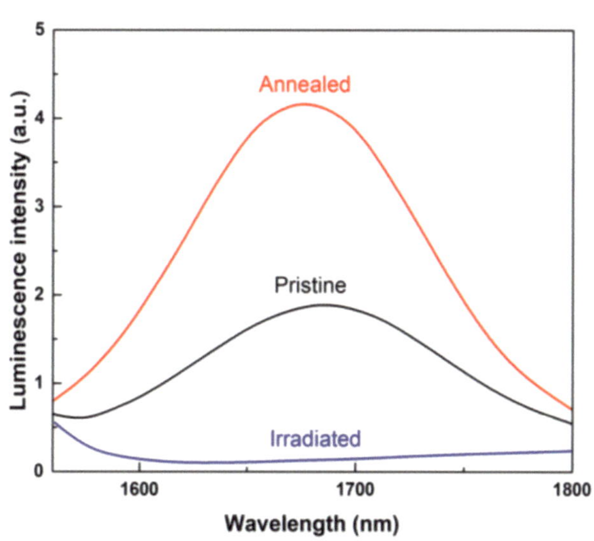

Figure 6.
Evolution of photoluminescence spectra of a BDF at different stages of the experiment: Pristine, irradiated, and annealed [19, 53].

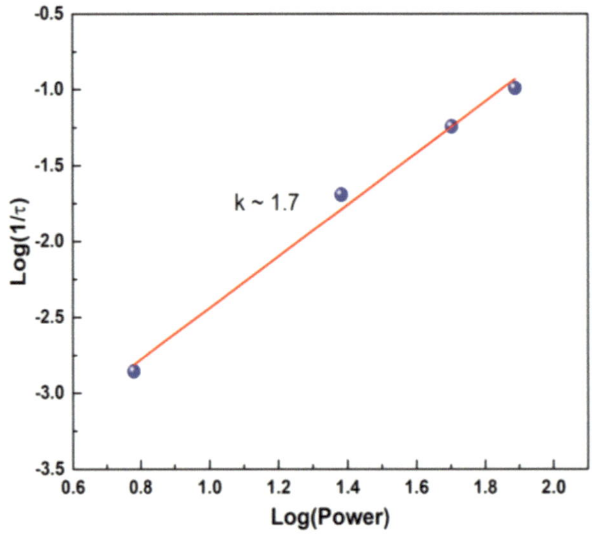

Figure 7.
Bleaching rate as a function of irradiation power under 532-nm irradiation (log-log scale) [48].

of BAC-Ge in BDFs has been observed when irradiated by 244, 407, 532, 639, 975, and 1460 nm [21, 45, 48, 51, 53]. In [48], the variation of BAC-Ge luminescence at 1700 nm in BDF under irradiation at different wavelengths with the irradiation intensity kept at ~0.5 MW/cm^2 has been demonstrated. Clearly, shorter irradiation wavelength provides higher photon energy, leading to stronger photobleaching of BAC-Ge. From this result, the photobleaching of BAC-Ge is almost unrelated to the resonant radiation wavelength. The mechanism of photobleaching of BAC-Ge will be discussed in detail in Section 4.4, which may be different from that of BAC-Si.

4.3 Temperature dependence

The temperature also has an influence upon the photobleaching of BAC-Ge. The BDF was irradiated under 532 nm with an intensity of 0.5 MW/cm^2 at room temperature (300 K) and liquid nitrogen temperature (77 K), respectively [48]. Seen from **Figure 8**, the photobleaching of BAC-Ge is suppressed significantly when the temperature is cooled down to 77 K. It is noticeable that the photoionization of the germanium-related oxygen deficient center (GeODC) decreases at low temperature [60]. Especially, it has already been confirmed that the BAC in bismuth-doped materials should be some cluster making up of Bi ion with the oxygen deficiency center (ODC) rather than Bi ions themselves [61]. Therefore, it is reasonable to deduce that the photobleaching of BAC-Ge may be related to the photoionization of GeODC. It was also found that the 1550-nm laser had no photobleaching effect upon BAC-Ge under 80°C but was able to bleach BAC-Ge when the temperature of BDF was elevated to hundreds of degrees [54]. Such combined effect of thermal treatment and laser irradiation on BAC-Al has also been observed in BEDF [52].

4.4 Photobleaching mechanism of BAC-Ge

It is known that the most convincing model of the nature of BAC is a Bi ion close to a structural defect, and the defect is most probably to be an ODC [62]. Furthermore, the mechanism of photobleaching of BAC-Ge is assumed to be the photoionization of GeODC. The photobleaching process of BAC-Ge induced by the destruction of GeODC by laser irradiation can be expressed as follows:

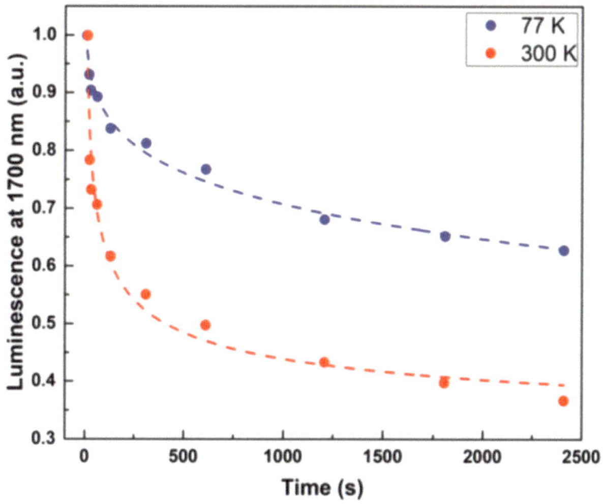

Figure 8.
Variation of BAC-Ge luminescence at 1700 nm in BDF irradiated by 532 nm at 300 and 77 K [48].

$$BAC \equiv \left(Bi^{+n} + ODC\right) \xrightarrow{2h\nu} Bi^{+n} + e^{-} + E^{'} center \quad (3)$$

The hypothesis is supported by the following facts: (1) the irradiation intensity dependence of bleaching rate demonstrates that photobleaching of BAC-Ge is likely to be a two-photon process [48]; (2) the GeODC can be photoionized under UV light irradiation [63]; and (3) the behavior of photobleaching of BAC-Ge is similar to the photoionization of GeODC at low temperature [48].

These results also suggest that the Bi ion adjacent to the ODC is the credible structure of BAC. According to this mechanism, it is believed that the number of the GeODC in the fiber should have an impact on the photobleaching of BAC-Ge. In this case, the doping concentration of Ge may affect the formation of GeODC, resulting in different photobleaching phenomena with various core compositions in [21]. This doping concentration dependence of photobleaching effect may also apply to the photobleaching of BAC-Si, BAC-Al, and BAC-P.

In addition, it is worth noting that the photobleaching of BAC-Si under 532-nm irradiation can be explained utilizing this hypothesis [45]. Therefore, more than one mechanism of photobleaching of BACs under different irradiation wavelengths should exist.

5. Photobleaching of BAC-Al

Compared with BAC-Si and BAC-Ge, the understanding of the fundamental structure of BAC-Al is still limited. A number of studies on photobleaching of BAC-Al have been taken to explore the nature of BAC-Al. It has been reported that the BAC-Al can be bleached with various irradiation conditions in bismuth/erbium co-doped aluminosilicate fibers [49]. **Figure 9** shows the absorption spectra of BEDF in the range from 650 to 750 nm before and after the irradiation. It is clear that the absorption of BAC-Al at ~700 nm decreases after the irradiation of 532, 633, 710, and 830 nm laser (for irradiation at 980 nm, the bleaching effect is not that obvious.). Unlike BAC-Si, after the irradiation, there is no obvious recovery behavior of BAC-Al at RT. Even the bleached BEDF is annealed at high temperatures; there remains no observable recovery of BAC-Al.

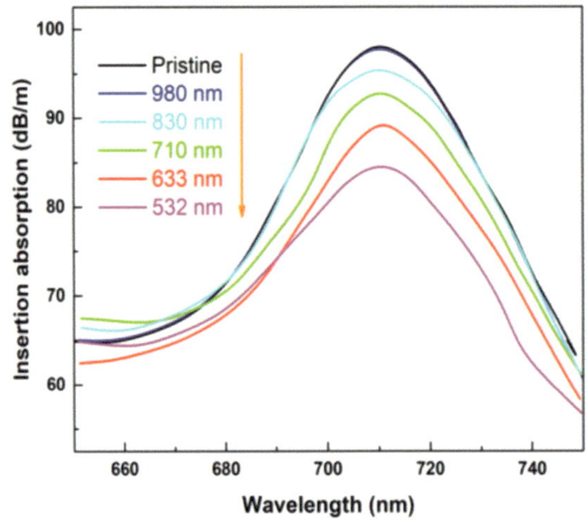

Figure 9.
Absorption of BEDF before and after irradiation at various wavelengths [49].

5.1 Irradiation intensity dependence

The photobleaching of BAC-Al also largely depends on the irradiation intensity [49, 52]. The intensity dependence of photobleaching of BAC-Al under 532-nm irradiation has been demonstrated in [49]. As irradiation intensity increases from 0.06 to 0.16 MW/cm^2, both the bleaching rate and bleaching ratio increase, indicating a faster and stronger photobleaching process. Moreover, the irradiation power dependence of bleaching rate in log-log scale shows a linear trend with a slope of ~1.8, as shown in **Figure 10**. The fitting slope close to 2 also demonstrates that the photobleaching of BAC-Al under 532-nm irradiation is likely to be a two-photon process.

5.2 Irradiation wavelength dependence

The BAC-Al can be bleached under irradiation at both resonant and nonresonant wavelengths. The different irradiation wavelengths (532, 633, 710, 830, and 980 nm) have been applied for the investigation of the photobleaching of Al-doped BEDF [49]. At RT, there is no obvious reduction of BAC-Al luminescence under 980-nm irradiation while the photobleaching of BAC-Al takes place under irradiation of all the other wavelengths. Especially, the stronger photobleaching of BAC-Al can be obtained by increasing the photon energy (reducing the irradiation wavelength) with a growth of the bleaching rate and a reduction of the unbleached ratio as shown in **Figure 11**. This irradiation wavelength dependence suggests that the Al-related ODC (AlODC) may take part in the photobleaching process instead of the Bi ion.

5.3 Temperature dependence

Similar to the BAC-Si and BAC-Ge, the photobleaching of BAC-Al is suppressed at low temperature. It has been reported that the bleaching ratio of BAC-Al under 0.12 MW/cm^2 532-nm irradiation is reduced from 10 to 5% as the temperature falls down from RT to 77 K [49]. In addition, the temperature aggravated photobleaching of BAC-Al has been observed in BEDF [52]. Under 0.34 MW/cm^2 980-nm irradiation, the BAC-Al luminescence at 1191 nm has little change at 293 K but decreases obviously at higher temperatures (423–623 K). The bleaching ratio of BAC-Al dramatically increases with the rising temperature, especially at 623 K, up to 35% of luminescence is bleached, as shown in **Figure 12**. The increasing temperature provides more

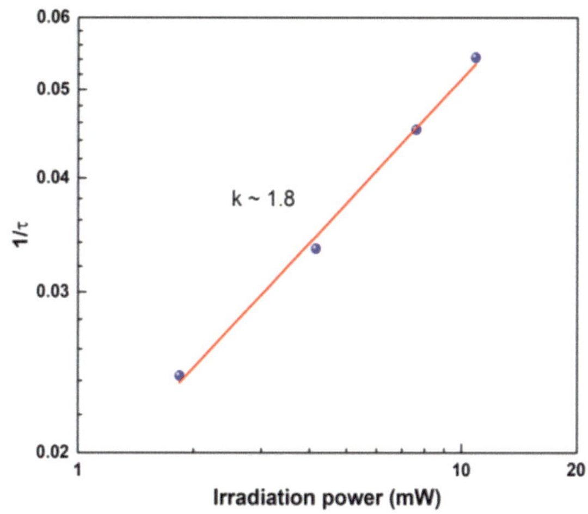

Figure 10.
The bleaching rate 1/τ of BAC-Al in BEDF under 532-nm irradiation versus irradiation power [49].

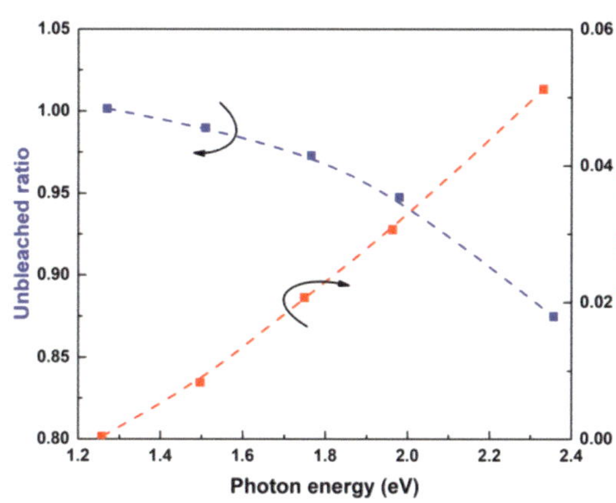

Figure 11.
Unbleached ratio and bleaching rate as a function of the photon energy [49].

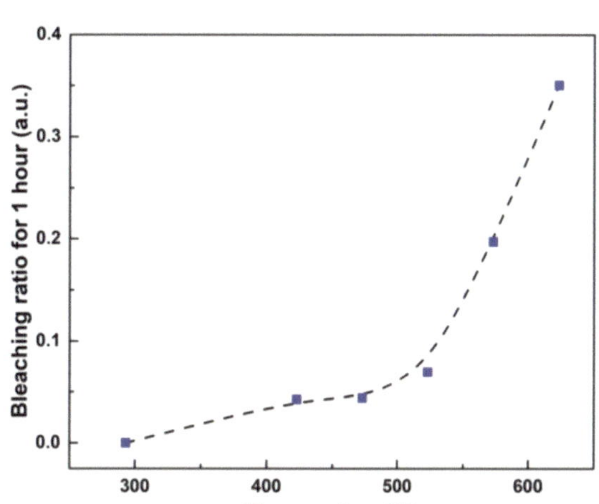

Figure 12.
Bleaching ratio of BAC-Al under 1 hour of 0.3 MW/cm² 980-nm irradiation as a function of the temperature [52].

thermal energy, making it possible for the electron to escape from the BAC-Al. The results imply the strong temperature dependence of photobleaching of BAC-Al, indicating the important role of thermal energy in the photobleaching process.

5.4 Photobleaching mechanism of BAC-Al

Considering that the BAC-Al can be bleached under irradiation of both resonant and nonresonant wavelengths, it is believed that the degradation of BAC-Al under irradiation is due to the photoinduced effects on AlODCs [49], which is much similar with BAC-Ge. Subsequently, the electron released from the AlODC is captured by the nearby defects, arousing the destruction of BAC-Al. In addition, the suppression of photobleaching of BAC-Al at the low temperature arising from the decreasing phonon-assisted rate for relaxation of AlODC, and the electron movement may support the view of participation of AlODC. However, unlike GeODC, the information of photo-ionization of AlODC is still limited, and there may exist more than one mechanism of photobleaching of BAC-Al. Therefore, the mechanism of photobleaching of BAC-Al still needs further investigation for deep understanding of the origin of BAC-Al.

6. Photobleaching of BAC-P

Only one photobleaching relevant study on the BAC-P has been reported so far. The study demonstrates that the luminescence of BAC-P at 1300 nm in BDF can be bleached under 1 MW/cm^2 407-nm irradiation [45]. The evolution of the luminescence of BAC-P is plotted as **Figure 13**, which is possibly linked with P-related ODC (PODC). Of course, more investigation on photobleaching of BAC-P needs to be taken in terms of irradiation intensity, irradiation wavelength, and temperature dependences to get in-depth knowledge of the photobleaching of BAC-P and its structure.

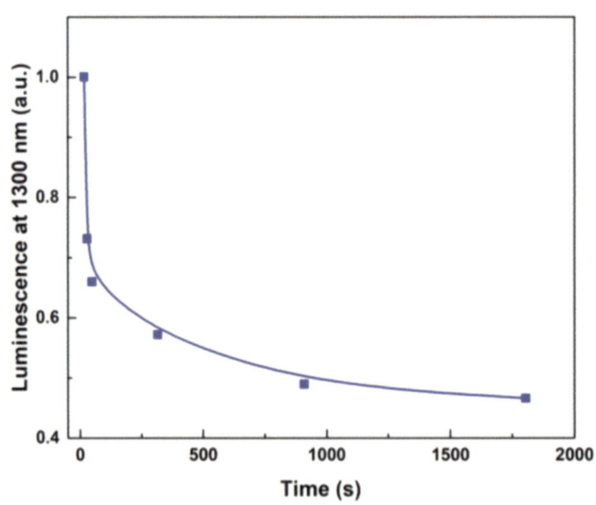

Figure 13.
Evolution of BAC-P luminescence at 1300 nm in BDF under 407-nm irradiation with an intensity of ~1 MW/cm^2 at room temperature [45].

7. Inductive analysis

The photobleaching effect exists in all four types of BACs. Since photobleaching effect varies case by case and is largely dependent on the irradiation conditions,

Fiber	Core composition	BAC type	Irradiation λ	Intensity	T	Time	r_B	Ref.
			nm	MW/cm^2	K	minutes		
BEDF	SiO_2-Al_2O_3-GeO_2-P_2O_5-Er_2O_3-Bi_2O_3 (~0.1 at%)	BAC-Si	830	0.004–0.36	300	360	0.2–0.57	[47]
BDF	10GeO_2– 90SiO_2: Bi	BAC-Si	407 532	1 1.5	300	30	0.85 0.8	[45]
BEDF	SiO_2-Al_2O_3-GeO_2-P_2O_5-Er_2O_3-Bi_2O_3 (~0.1 at%)	BAC-Si	980, 830, 710, 1380	0.36 0.005	300	—	0–0.66 0.06	[47]
BEDF	SiO_2-Al_2O_3-GeO_2-P_2O_5-Er_2O_3-Bi_2O_3 (~0.1 at%)	BAC-Si	830	0.36	77–673	—	0.34–0.66	[47]
BDF	50GeO_2–50SiO_2: Bi (<0.1 mol%)	BAC-Ge	532	0.6–1.2	300	60	—	[21]
BDF	95GeO_2–5SiO_2: Bi (<0.1 mol%)	BAC-Ge	532	0.6–1.2	300	60	—	[21]
BDF	50GeO_2–50SiO_2: Bi (<0.1 mol%)	BAC-Ge	244	—	300	—	1	[51]
BDF	50GeO_2–50SiO_2: Bi (~100 ppm)	BAC-Ge	1460, 975, 639, 532, 407	0.5	300	60	0.05–0.98	[48]
BDF	50GeO_2–50SiO_2: Bi (~100 ppm)	BAC-Ge	532	0.5	77, 300	40	0.35, 0.6	[48]
BEDF	SiO_2-Al_2O_3-Er_2O_3-Bi_2O_3 (~0.1 at%)	BAC-Al	532	0.06–0.16	300	55	0.05–0.15	[49]
BEDF	SiO_2-Al_2O_3-Er_2O_3-Bi_2O_3 (~0.1 at%)	BAC-Al	980, 830, 710, 633, 532	0.12	300	55	0–0.1	[49]
BEDF	SiO_2-Al_2O_3-GeO_2-P_2O_5-Er_2O_3-Bi_2O_3 (~0.1 at%)	BAC-Al	980	0.34	293–623	60	0–0.35	[52]
BDF	5P_2O_5–95SiO_2: Bi	BAC-P	407	1	300	30	0.53	[45]

Note: T, temperature of the fiber when it is irradiated by the laser.

Table 2.
Summary of photobleaching of BACs in BDFs and BEDFs.

here, **Table 2** summarizes a series of photobleaching of BACs observed in BDFs and BEDFs, along with their fiber compositions, BAC type, and irradiation conditions. The samples are doped with different compositions, such as Si, Ge, Al, P, Bi, and Er. The irradiation wavelengths are from 244 to 1460 nm, and the irradiation power varies from 0.005 to 1.5 MW/cm^2. The exposure temperature of the sample is in the range of 77–673 K.

In general, the bleaching ratio increases with irradiation intensity. Higher irradiation intensity provides more photons, leading to stronger photobleaching effect. It is worth noting that for both BAC-Si and BAC-Al, the bleaching ratio tends to saturate as the irradiation intensity increases [47, 52]. More interestingly, the photobleaching effect could happen under irradiation of some wavelengths even the irradiation intensity is quite small. However, for some irradiation wavelengths, even the irradiation intensity is large, there is still no obvious photobleaching phenomenon. For example, the luminescence of BAC-Si decays 20% after 830-nm irradiation with an intensity of 0.005 MW/cm^2 but has no change when exposed to 0.36 MW/cm^2 980-nm irradiation [47]. This indicates that the irradiation wavelength affects more on the photobleaching than the irradiation intensity.

The photobleaching effect significantly depends on the irradiation wavelength. As for BAC-Ge and BAC-Al, the photobleaching effect could happen by the irradiations at both resonant and nonresonant wavelengths, and larger bleaching ratio can be achieved by shorter irradiation wavelength [48, 49]. However, as for BAC-Si, only the wavelengths that are able to excite BAC-Si to the upper level could lead to the photobleaching [47]. According to **Table 2**, it is worth noting that irradiation wavelengths that can cause the photobleaching effect are always shorter than the luminescence peak wavelength of BACs. Therefore, it is supposed that the premise of photobleaching is that the photon energy for the photobleaching is larger than the excitation energy between the ground state and the first excited state of BACs.

Higher temperature evidently accelerates the electron movement rate. It is believed that the photobleaching of BACs is due to the electron escape. Therefore, the photobleaching of BACs becomes stronger at high temperatures, as demonstrated in [47, 48, 52, 54]. It is remarkable that for some irradiation wavelengths, the luminescence has little change at room temperature but is bleached significantly at higher temperatures [52, 54]. The increasing temperature provides enough thermal energy and assists the electron to flee from the BACs; however, the phonon energy is not sufficient for electron escape at room temperature.

8. Summary

Since the first observation of NIR luminescence in bismuth-doped glass, the bismuth-doped materials have attracted great attention due to their ultra-broadband luminescence. Especially, the bismuth-doped and bismuth/erbium co-doped optical fibers have been developed for the potential applications as optical amplifier and fiber laser. A number of researches have been taken focusing on the photostability of bismuth active center (BAC) in these BDFs/BEDFs. The results have demonstrated that the laser radiation can evidently cause the photobleaching of all types of BACs (BAC-Si, BAC-Ge, BAC-Al, and BAC-P), leading to the change of optical characteristics of the fiber. For BAC-Si, BAC-Ge, and BAC-Al, the photobleaching is much dependent upon the irradiation intensity, irradiation wavelength, and temperature. The recovery behavior of bleached BAC-Si and BAC-Ge can be achieved with the aid of phonon-assisted relaxation after the irradiation. It is noted that the BAC-Ge and BAC-Al can be bleached under irradiation at both resonant and nonresonant wavelengths; however, the photobleaching of BAC-Si can only happen when the irradiation photon is able to excite the BAC-Si to the upper energy level. These differences indicate that the photobleaching of BACs is driven by multiple possible mechanisms. In bismuth-doped germanosilicate fiber, the possible mechanism of the photobleaching effect is the photoionization of GeODC, which participate in the formation of BAC-Ge. In addition, the fabrication process, material compositions, treatment conditions, as well as doping concentration of Si/Ge/Al/P have a great impact on the formation of their related ODCs (GeODC, AlODC, PODC, etc.) and ultimately affect the photobleaching of related BACs. All these investigations on photobleaching of BACs in these BDFs/BEDFs not only provide their photostability information but also give an insight to reveal the fundamental structure of BACs, which can be utilized to control the BACs in BDFs and BEDFs for the practical applications as an optical amplifier and fiber laser.

Acknowledgements

The authors are thankful for the support of National Natural Science Foundation of China (61520106014 and 61675032), Science and Technology Commission of Shanghai Municipality, China (SKLSFO2018-02) and 111 Project (D20031) and wish to express their thanks to other collaborators for their contributions.

Author details

Bowen Zhang[1], Mingjie Ding[1], Shuen Wei[1], Binbin Yan[2], Gang-Ding Peng[1], Yanhua Luo[1*] and Jianxiang Wen[3]

1 Photonics and Optical Communications, School of Electrical Engineering and Telecommunications, University of New South Wales, Sydney, NSW, Australia

2 State Key Laboratory of Information Photonics and Optical Communications, Beijing University of Posts and Telecommunications, Beijing, China

3 Key laboratory of Specialty Fiber Optics and Optical Access Networks, Joint International Research Laboratory of Specialty Fiber Optics and Advanced Communication, Shanghai University, Shanghai, China

*Address all correspondence to: yanhua.luo1@unsw.edu.au

References

[1] Murata K, Fujimoto Y, Kanabe T, Fujita H, Nakatsuka M. Bi-doped SiO_2 as a new laser material for an intense laser. Fusion Engineering and Design. 1999;**44**(1-4):437-439

[2] Denker B, Galagan B, Osiko V, Sverchkov S, Dianov E. Luminescent properties of Bi-doped boro-alumino-phosphate glasses. Applied Physics B. 2007;**87**(1):135-137

[3] Peng M, Dong G, Wondraczek L, Zhang L, Zhang N, Qiu J. Discussion on the origin of NIR emission from Bi-doped materials. Journal of Non-Crystalline Solids. 2011;**357**(11-13):2241-2245

[4] Peng M, Zollfrank C, Wondraczek L. Origin of broad NIR photoluminescence in bismuthate glass and Bi-doped glasses at room temperature. Journal of Physics. Condensed Matter. 2009;**21**(28):285106

[5] Ruan J, Su L, Qiu J, Chen D, Xu J. Bi-doped BaF_2 crystal for broadband near-infrared light source. Optics Express. 2009;**17**(7):5163-5169

[6] Suzuki T, Ohishi Y. Ultrabroadband near-infrared emission from Bi-doped Li_2O–Al_2O_3–SiO_2 glass. Applied Physics Letters. 2006;**88**(19):191912

[7] Zhou S, Dong H, Zeng H, Feng G, Yang H, Zhu B, et al. Broadband optical amplification in Bi-doped germanium silicate glass. Applied Physics Letters. 2007;**91**(6):061919

[8] Chu Y, Hu Q, Zhang Y, Gao Z, Fang Z, Liu L, et al. Topological engineering of photoluminescence properties of bismuth-or erbium-doped phosphosilicate glass of arbitrary P_2O_5 to SiO_2 ratio. Advanced Optical Materials. 2018;**6**(13):1800024

[9] Ahmad H, Shahi S, Harun SW. Bismuth-based erbium-doped fiber as a gain medium for L-band amplification and brillouin fiber laser. Laser Physics. 2010;**20**(3):716-719

[10] Bufetov I, Dianov E. Bi-doped fiber lasers. Laser Physics Letters. 2009;**6**(7):487

[11] Bufetov IA, Firstov SV, Khopin VF, Medvedkov OI, Guryanov AN, Dianov EM. Bi-doped fiber lasers and amplifiers for a spectral region of 1300-1470 nm. Optics Letters. 2008;**33**(19):2227-2229

[12] Bufetov IA, Melkumov MA, Firstov SV, Riumkin KE, Shubin AV, Khopin VF, et al. Bi-doped optical fibers and fiber lasers. IEEE Journal of Selected Topics in Quantum Electronics. 2014;**20**(5):111-125

[13] Dvoyrin V, Medvedkov O, Mashinsky V, Umnikov A, Guryanov A, Dianov E. Optical amplification in 1430-1495 nm range and laser action in Bi-doped fibers. Optics Express. 2008;**16**(21):16971-16976

[14] Luo Y, Wen J, Zhang J, Canning J, Peng G-D. Bismuth and erbium codoped optical fiber with ultrabroadband luminescence across O-, E-, S-, C-, and L-bands. Optics Letters. 2012;**37**(16):3447-3449

[15] Luo Y, Yan B, Zhang J, Wen J, He J, Peng G-D. Development of Bi/Er co-doped optical fibers for ultra-broadband photonic applications. Frontiers of Optoelectronics. 2018;**11**(1):37-52

[16] Wei S, Ding M, Fan D, Luo Y, Wen J, Peng G-D. Effects of post treatments on bismuth-doped and bismuth/erbium co-doped optical fibres. Bismuth: Advanced Applications and Defects Characterization. 2018;**155**

[17] Firstov S, Khopin V, Bufetov I, Firstova E, Guryanov A,

Dianov E. Combined excitation-emission spectroscopy of bismuth active centers in optical fibers. Optics Express. 2011;**19**(20):19551-19561

[18] Sathi ZM, Zhang J, Luo Y, Canning J, Peng G. Spectral properties and role of aluminium-related bismuth active Centre (BAC-Al) in bismuth and erbium co-doped fibres. Optical Materials Express. 2015;**5**(5):1195-1209

[19] Firstov SV, Alyshev SV, Riumkin KE, Khegai AM, Kharakhordin AV, Melkumov MA, et al. Laser-active fibers doped with bismuth for a wavelength region of 1.6-1.8 μm. IEEE Journal of Selected Topics in Quantum Electronics. 2018;**24**(5):1-15

[20] Wen J, Liu W, Dong Y, Luo Y, Peng G-D, Chen N, et al. Radiation-induced photoluminescence enhancement of Bi/Al-codoped silica optical fibers via atomic layer deposition. Optics Express. 2015;**23**(22):29004-29013

[21] Firstov S, Alyshev S, Khopin V, Melkumov M, Guryanov A, Dianov E. Photobleaching effect in bismuth-doped germanosilicate fibers. Optics Express. 2015;**23**(15):19226-19233

[22] Firstov SV, Alyshev SV, Khopin VF, Kharakhordin AV, Lobanov AS, Firstova EG, et al. Effect of heat treatment parameters on the optical properties of bismuth-doped GeO_2: SiO_2 glass fibers. Optical Materials Express. 2019;**9**(5):2165-2174

[23] Wei S, Luo Y, Ding M, Cai F, Xiao G, Fan D, et al. Thermal effect on attenuation and luminescence of Bi/Er co-doped fiber. IEEE Photonics Technology Letters. 2016;**29**(1):43-46

[24] Wei S, Luo Y, Fan D, Xiao G, Chu Y, Zhang B, et al. BAC activation by thermal quenching in bismuth/erbium codoped fiber. Optics Letters. 2019;**44**(7):1872-1875

[25] Zhang B, Wei S, Khan MTA, Luo Y, Peng G-D. Dynamics study of thermal activation of BAC-Si in bismuth/erbium-codoped optical fiber. Optics Letters. 2020;**45**(2):571-574

[26] Song L, Hennink E, Young IT, Tanke HJ. Photobleaching kinetics of fluorescein in quantitative fluorescence microscopy. Biophysical Journal. 1995;**68**(6):2588-2600

[27] Greenbaum L, Rothmann C, Lavie R, Malik Z. Green fluorescent protein photobleaching: A model for protein damage by endogenous and exogenous singlet oxygen. Biological Chemistry. 2000;**381**(12):1251-1258

[28] Krebs J, Segal C, Yen W, Happek U. Photo bleaching in Sm^{2+} selectively doped epitaxial CaF_2 films. Journal of Luminescence. 1997;**72**:220-221

[29] Pollnau M, Hardman P, Clarkson W, Hanna D. Upconversion, lifetime quenching, and ground-state bleaching in Nd^{3+}: $LiYF_4$. Optics Communication. 1998;**147**(1-3):203-211

[30] Takeshita S, Ogata H, Isobe T, Sawayama T, Niikura S. Effects of citrate additive on transparency and photostability properties of YVO_4: Bi^{3+}, Eu^{3+} nanophosphor. Journal of the Electrochemical Society. 2010;**157**(3):J74-J80

[31] Busch G, Greve K, Olson G, Jones R, Rentzepis P. Photobleaching and recovery times of the mode-locking dye DODCI. Chemical Physics Letters. 1975;**33**(3):412-416

[32] Oster G, Wotherspoon N. Photobleaching and photorecovery of dyes. The Journal of Chemical Physics. 1954;**22**(1):157-158

[33] Saylor J. Photobleaching of disodium fluorescein in water. Experiments in Fluids. 1995;**18**(6):445-447

[34] Mills A, Wang J. Photobleaching of methylene blue sensitised by TiO_2: An ambiguous system? Journal of Photochemistry and Photobiology, A: Chemistry. 1999;**127**(1-3):123-134

[35] Henderson JN, H-w A, Campbell RE, Remington SJ. Structural basis for reversible photobleaching of a green fluorescent protein homologue. Proceedings of the National Academy of Sciences. 2007;**104**(16):6672-6677

[36] Sinnecker D, Voigt P, Hellwig N, Schaefer M. Reversible photobleaching of enhanced green fluorescent proteins. Biochemistry. 2005;**44**(18):7085-7094

[37] White J, Stelzer E. Photobleaching GFP reveals protein dynamics inside live cells. Trends in Cell Biology. 1999;**9**(2):61-65

[38] Chávez AG, Kir'Yanov A, Barmenkov YO, Il'Ichev N. Reversible photo-darkening and resonant photo-bleaching of ytterbium-doped silica fiber at in-core 977-nm and 543-nm irradiation. Laser Physics Letters. 2007;**4**(10):734

[39] Manek-Hönninger I, Boullet J, Cardinal T, Guillen F, Ermeneux S, Podgorski M, et al. Photodarkening and photobleaching of an ytterbium-doped silica double-clad LMA fiber. Optics Express. 2007;**15**(4):1606-1611

[40] Gebavi H, Taccheo S, Tregoat D, Monteville A, Robin T. Photobleaching of photodarkening in ytterbium doped aluminosilicate fibers with 633 nm irradiation. Optical Materials Express. 2012;**2**(9):1286-1291

[41] Laperle P, Chandonnet A, Vallée R. Photobleaching of thulium-doped ZBLAN fibers with visible light. Optics Letters. 1997;**22**(3):178-180

[42] Savelyev E, Butov O, Yapaskurt V, Golant K. Near-infrared luminescence of bismuth in silica-based glasses with different additives. Journal of Communications Technology and Electronics. 2018;**63**(12):1458-1468

[43] Vtyurina D, Romanov A, Kuznetsov M, Fattakhova Z, Khaula E, Lisitskii I, et al. Optical properties of bismuth-doped $TlCdCl_3$ crystal. Russian Journal of Physical Chemistry B. 2016;**10**(1):1-4

[44] Wang X, Xu S, Yang Z, Peng M. Ultra-broadband red to NIR photoemission from multiple bismuth centers in $Sr_2B_5O_9Cl$: Bi crystal. Optics Letters. 2019;**44**(19):4821-4824

[45] Firstov SV, Alyshev SV, Kharakhordin AV, Riumkin KE, Dianov EM. Laser-induced bleaching and thermo-stimulated recovery of luminescent centers in bismuth-doped optical fibers. Optical Materials Express. 2017;**7**(9):3422-3432

[46] Ding M, Wei S, Luo Y, Peng G-D. Reversible photo-bleaching effect in a bismuth/erbium co-doped optical fiber under 830 nm irradiation. Optics Letters. 2016;**41**(20):4688-4691

[47] Ding M, Fang J, Luo Y, Wang W, Peng G-D. Photo-bleaching mechanism of the BAC-Si in bismuth/erbium co-doped optical fibers. Optics Letters. 2017;**42**(24):5222-5225

[48] Firstov SV, Alyshev SV, Firstova EG, Melkumov MA, Khegay AM, Khopin VF, et al. Dependence of the photobleaching on laser radiation wavelength in bismuth-doped germanosilicate fibers. Journal of Luminescence. 2017;**182**:87-90

[49] Zhao Q, Luo Y, Tian Y, Peng G-D. Pump wavelength dependence and thermal effect of photobleaching of BAC-Al in bismuth/erbium codoped aluminosilicate fibers. Optics Letters. 2018;**43**(19):4739-4742

[50] Ding M, Luo Y, Wen J, Peng G-D, editors. Dynamic behavior of pump

light radiation induced photo-bleaching effect on BAC-Si in bismuth/erbium co-doped optical fibers. In: Fiber Lasers XV: Technology and Systems. Washington, D.C., USA: International Society for Optics and Photonics; 2018

[51] Firstov S, Alyshev S, Melkumov M, Riumkin K, Shubin A, Dianov E. Bismuth-doped optical fibers and fiber lasers for a spectral region of 1600-1800 nm. Optics Letters. 2014;**39**(24):6927-6930

[52] Zhang B, Luo Y, Ding M, Wei S, Fan D, Xiao G, et al. Thermally aggravated photo-bleaching of BAC-Al in bismuth/erbium co-doped optical fiber. Optics Letters. 2019;**44**(19):4829-4832

[53] Firstov S, Firstova E, Alyshev S, Khopin V, Riumkin K, Melkumov M, et al. Recovery of IR luminescence in photobleached bismuth-doped fibers by thermal annealing. Laser Physics. 2016;**26**(8):084007

[54] Alyshev S, Kharakhordin A, Firstova E, Khegai A, Melkumov M, Khopin V, et al. Photostability of laser-active centers in bismuth-doped GeO_2–SiO_2 glass fibers under pumping at 1550 nm. Optics Express. 2019;**27**(22):31542-31552

[55] Johnston D. Stretched exponential relaxation arising from a continuous sum of exponential decays. Physical Review B. 2006;**74**(18):184430

[56] Zhang B, Wei S, Chu Y, Talal M, Fu X, Yan B, et al., editors. Thermal quenching effect on BAC-P in bismuth/erbium co-doped optical fibre. In: Asia Communications and Photonics Conference. Washington, D.C., USA: Optical Society of America; 2019

[57] Zhao Q, Luo Y, Hao Q, Peng G-D. Effect of thermal treatment parameters on the spectral characteristics of BAC-Al in bismuth/erbium-codoped aluminosilicate fibers. Optics Letters. 2019;**44**(18):4594-4597

[58] Xu H, Yan B, Lin J, Luo Y, Lu P, Wang K, et al. Effects of quenching and cooling upon near infrared luminescence of Bi/Er co-doped optical fiber. Optical Materials Express. 2019;**9**(7):3156-3168

[59] Kharakhordin A, Alyshev S, Firstova E, Khegai A, Melkumov M, Khopin V, et al. Analysis of thermally activated processes in bismuth-doped GeO_2-SiO_2 glass fibers using the demarcation energy concept. Optical Materials Express. 2019;**9**(11):4239-4246

[60] Hosono H, Abe Y, Kinser DL, Weeks RA, Muta K, Kawazoe H. Nature and origin of the 5-eV band in SiO_2: GeO_2 glasses. Physical Review B. 1992;**46**(18):11445

[61] Dianov EM. Nature of Bi-related near IR active centers in glasses: State of the art and first reliable results. Laser Physics Letters. 2015;**12**(9):095106

[62] Dianov EM. On the nature of near-IR emitting Bi centres in glass. Quantum Electronics. 2010;**40**(4):283

[63] Vasil'ev SA, Dianov EM, Koltashev VVE, Marchenko VM, Mashinsky VM, Medvedkov OI, et al. Photoinduced changes in the Raman spectra of germanosilicate optical fibres. Quantum Electronics. 1998;**28**(4):330

Chapter 4

Radiation Effect on Optical Properties of Bi-Related Materials Co-Doped Silica Optical Fibers

Jianxiang Wen, Ying Wan, Yanhua Dong, Yi Huang, Yanhua Luo, Gang-Ding Peng, Fufei Pang and Tingyun Wang

Abstract

Three kinds of Bi-related materials co-doped silica optical fibers (BRDFs), including Bi/Al, Bi/Pb, and Bi/Er co-doped fibers, were fabricated using atomic layer deposition (ALD) and modified chemical vapor deposition (MCVD). Then, the effect of irradiation on the optical properties of BRDFs was investigated. The experimental results showed that the fluorescence intensity, the fluorescence lifetime of BRDFs at the 1150 nm band, increased significantly with low-dose treatment, whereas it decreased with a further increase in the radiation dose. In addition, the merit M_α values of the BRDFs, a ratio of useful pump absorption to total pump absorption, decreased with an increase of the radiation doses. The Verdet constants of different doped fibers increased up to saturation level with increases in the radiation dose. However, for a Bi-doped fiber, its Verdet constant decreased and the direction of Faraday's rotation changed under low-dose radiation treatment. In addition, the Verdet constant increase of the Bi-doped silica fiber was much faster than that of other single mode fiber (SMF) and Pb-doped silica fibers treated with high-dose radiation. All of these findings are of great significance for the study of the optical properties of BRDFs.

Keywords: Bi-related materials, optical properties, gamma-ray radiation, silica optical fiber

1. Background and meaning

Bismuth oxide is an important material with many promising applications [1, 2]. In particular, bismuth-doped optical fibers, as a promising active medium for amplifying and lasing in the 1.1–1.8 μm range [3, 4], have been extensively studied, ever since their broadband near-infrared (NIR) fluorescence properties were first reported by Fujimoto et al. [1]. Thereafter, an amplification at the 1300 nm band in Bi-doped silica glass was realized [2] and an optical amplifier and fiber laser were achieved [4–6]. Previous investigations have demonstrated that the valence state of Bi ions varies in glass materials [7–11]. However, the valence conversion mechanism of Bi-related materials for silica optical fibers remains unclear.

In addition, the effects of radiation on the fluorescence properties of Bi-doped glass or optical fibers have been previously studied [12–18]. In Refs. [16–18], the

IntechOpen

radiation-induced photoluminescence (PL) effect of Bi-doped silica optical fibers was investigated and the relationships between the radiation-induced optical properties and defect centers in Bi-doped silica fibers (BDFs) were reported. Moreover, the fluorescence intensity was enhanced by UV irradiation [11, 12]. Shen et al. [14] also reported fluorescence enhancement from exposing Bi-doped borosilicate glass to a radiation environment. The photo-bleaching effect on Bi-doped glass fiber with a 532-nm laser treatment was studied [15]. These results provide deeper insight into the nature and formation mechanism of PL [19]. Furthermore, the magneto-optical properties of the Bi-doped silica fibers were studied before and after irradiation, and radiation-induced magneto-optical phenomena were found [20–22]. Finally, thermal effects on the luminescence properties of Bi co-doped silica fibers were studied [23–25].

However, the nature of the NIR fluorescence properties in Bi-doped glass or silica optical fibers is still unclear. Although many studies have reported the luminescent properties of Bi co-doped fibers, there are few reports on the effect of irradiation on the optical properties of Bi-related co-doped silica optical fibers.

In this chapter, three kinds of Bi-related co-doped silica optical fibers, including Bi/Al, Bi/Pb, and Bi/Er co-doped fibers, are fabricated using atomic layer deposition (ALD) combined with a modified chemical vapor deposition (MCVD) process. The optical properties of bi-related materials co-doped silica optical fibers (BRDFs) that are influenced by irradiation are investigated, including luminescence, lifetime decay, magnetic-optical, and unsaturable absorption, and the changes in these optical properties are compared.

2. Fabrication of Bi-related co-doped silica optical fibers

Currently, the fabrication technologies of different doped fibers such as rare earth-doped fibers mainly use a solution-doping chemical vapor deposition technique. However, the technology lacks uniformity and consistency, and doping materials are easily volatilized and form clusters in a high-temperature environment, which limits the excellent performance of the fabricated doped fibers. Recently, a novel doping method, ALD technology, has been developed. It is not only an advanced deposition technique [5, 26–30] that allows for ultrasmall dopants of a few nanometers to be deposited in a precisely controlled way but also a chemical vapor deposition technique based on the sequential use of self-terminating gas–solid reactions. In particular, the novel technology involves a self-limiting surface reaction, whose advantages include a low-temperature process, good uniformity, favorable dispersibility, high doping concentration, and wide range of materials used. To date, there have been only a few reports [30–34] regarding the preparation of rare earth optical fibers by ALD.

ALD technology typically involves a cycle of four steps that is repeated as many times as necessary to achieve the required doping concentrations. As an example, we perform ALD on Al_2O_3, using $Al_2(CH_3)_3$ (Trimethylaluminum, TMA) and H_2O as the reactants. The detailed deposition process is shown in **Figure 1**. **Step 1:** A pulse precursor vapor of TMA reacts with the inner surface of the substrate tube. With the optimized choice of precursors and reaction conditions, the reaction of this step is self-limiting. **Step 2:** Purging all residual precursors and reaction products. **Step 3:** Low damage by remote exposure of the surface to reactive oxygen radicals, where these radicals oxidize the inner surface and remove surface ligands. This reaction is also self-limiting because of the limited number of surface ligands. **Step 4:** Reaction products are purged from the chamber. Only Step 3 varies, between H_2O and O_2 plasma by thermal processing. As each cycle of the ALD process deposits

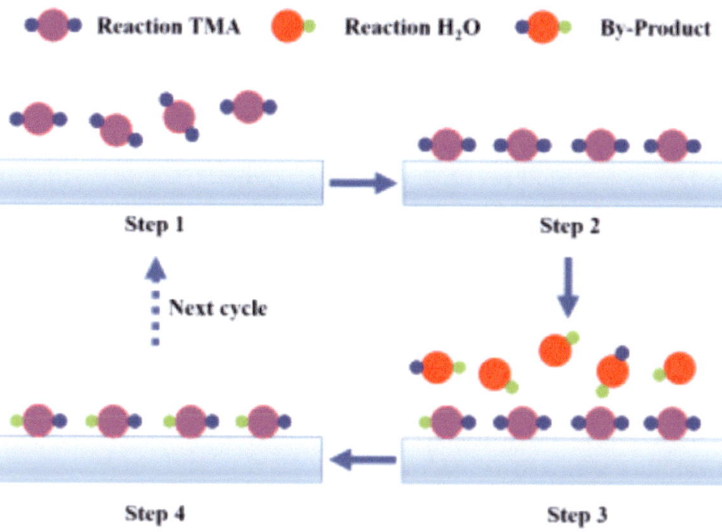

Figure 1.
Specific process of ALD technology deposition.

a layer of sub-angstrom thickness, the atomic scale ranges of deposition can be controlled through the process.

The reaction of $X(thd)_3$ (X: metal ions, such as Bi, Pb, and Er; thd: 2,2,6,6-tetramethyl-3,5-heptanedionato) and H_2O can be described by Eqs. (1)–(3) [35]. The whole reaction can be written as follows:

$$2X(thd)_3 + 3H_2O \rightarrow X_2O_3 + 6H - thd \tag{1}$$

which involves two processes: process A in Eq. (2) is the hydroxyl on silicon reacting with the X source to obtain Si-O-$X(thd)_2$; process B is obtaining Si-O-$X(OH)_2$ by the reaction in Eq. (3) of H_2O and Si-O-$X(thd)_2$ with the termination of —OH groups. On repeating the ABAB (A and B represent different reaction processes, respectively) operations, an X-doped layer with the desired thickness is obtained. Similarly, Al_2O_3 can be deposited using these following analogous reactions.

$$A: \; Si - OH + X(thd)_3 \rightarrow Si - O - X(thd)_2 + H - thd \tag{2}$$

$$B: \; Si - O - X(thd)_2 + 2H_2O \rightarrow Si - O - X(OH)_2 + 2H - thd \tag{3}$$

The fabrication process of the BRDFs can be divided into four steps, as shown in **Figure 2**. First, a porous soot layer is deposited inside the silica substrate tube using

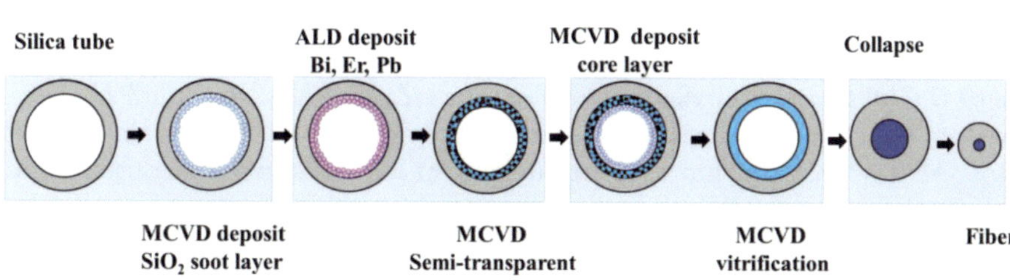

Figure 2.
Fabrication process of the BRDFs based on ALD + MCVD technology.

the MCVD method. In this process, chemical reactions in the gas phase generate a fine soot of silica that coats the inner surface of the substrate tube, which is then sintered into a semi-clear soot layer. Second, Bi, Pb, or Er ions are introduced on the surface of the porous soot layer using the ALD technique (TFS-200, Beneq, Finland). This results in the formation of bismuth oxide, lead oxide, and erbium oxide with the precursors of bis (2, 2, 6, 6-tetra-methyl-3, 5-heptanedionato) bismuth (III) ($Bi(thd)_3$), bis (2, 2, 6, 6-tetra-methyl-3, 5-heptanedionato) lead (III) ($(Pb(thd))_3$, and bis (2, 2, 6, 6-tetra-methyl-3, 5-heptanedionato) erbium (III) ($(Er(thd))_3$ (supplied by Shanghai J&K Scientific Ltd), respectively. They mainly react with water or ozone to form the metal oxidation layer, the O_3 that originated from the O_2. Third, germanium oxide is doped into the fiber preform core by the MCVD process, and then a Bi-related co-doped optical fiber preforms with a Ge-doped higher index core that is formed by collapsing on an MCVD lathe heated by a high-temperature oxyhydrogen flame. Finally, the preform is drawn into a doped optical fiber with a Bi-related material.

3. Effect of radiation on optical properties

For optical fiber material, a perfect structure is visualized as a co-doped ion random network of SiO_4 tetrahedrons joined at the corner, and different ions are doped into irregular vitreous silica, forming a stable network structure [36]. It is important to accumulate further knowledge regarding the influence of radiation on optical fiber materials, including material network structures, defect centers, and optical properties. Radiation as an effective method can induce changes in the optical properties of materials. It mainly involves the process of high-energy particles interacting with fiber materials, including the photoelectric effect, the Compton effect, the electron pair effect, and more. For BRDFs, irradiation significantly improves their optical properties, which mainly accounts for the variation in the valence states of Bi (Bi^{5+}, Bi^{2+}, Bi^{+}, Bi^{0}, defect centers, Bi clusters, $Bi^{2-}{}_2$ dimers, or Bi atoms). Here, gamma rays are selected as the irradiation source, mainly due to their short wavelength and strong penetrating ability. The effects of gamma ray irradiation on the optical properties of BRDFs, including Bi/Al co-doped silica fibers (BADFs), Bi/Er co-doped silica fibers (BEDFs), and Bi/Pb co-doped silica fibers (BPDFs), are investigated.

3.1 Radiation effect on luminescence characteristics

The radiation-induced PL properties of BADFs were investigated in [19]. The PL spectra in the inset of **Figure 3** reveal two emission bands at approximately ~1150 and ~ 1410 nm, corresponding to the aluminum-related Bi active center (BAC-Al) and the silicon-related Bi active center (BAC-Si), respectively. **Figure 3** illustrates that the fluorescence intensities of BAC-Al increased by 0.73, 2.25, and 1.35 dB at 1150 nm with 1.0, 2.0, and 3.0 kGy of irradiation, respectively. The fluorescence intensities of BAC-Al in the BADF samples increased with the increase in radiation dose (0–2.0 kGy) and then decreased when the radiation dose exceeded 2 kGy. Moreover, the change in the fluorescence intensity of BAC-Si trended similar to that of BAC-Al; however, the fluorescence intensity of BAC-Si increased considerably more. Furthermore, the fluorescence intensity of BAC-Si was approximately four times stronger than that of the unirradiated fiber sample.

For BEDF, five BEDF samples were irradiated with cumulative doses of approximately 0.3, 0.5, 0.8, 1.5, and 3.0 kGy at room temperature. The radiation dose rate was 800 Gy/h. Under excitation at 980 nm (pump power is 1.8 mW),

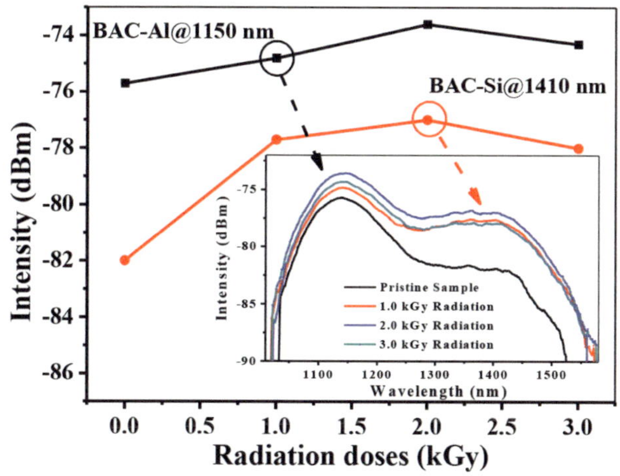

Figure 3.
Fluorescence intensity of BAC-Al and BAC-Si as a function of radiation dose; inset are the PL spectra of the BADF samples before and after γ-ray irradiation.

the fluorescence spectra of BEDF samples were measured, as shown in **Figure 4**. For BAC-Al, as the radiation dose was increased, the fluorescence intensity first increased and then decreased. With a 0.3 kGy dose of irradiation, the fluorescence intensity of BAC-Al in the BEDF sample is slightly higher than that of the pristine fiber, as shown in **Figure 4(a)**. However, when the radiation dose was less than

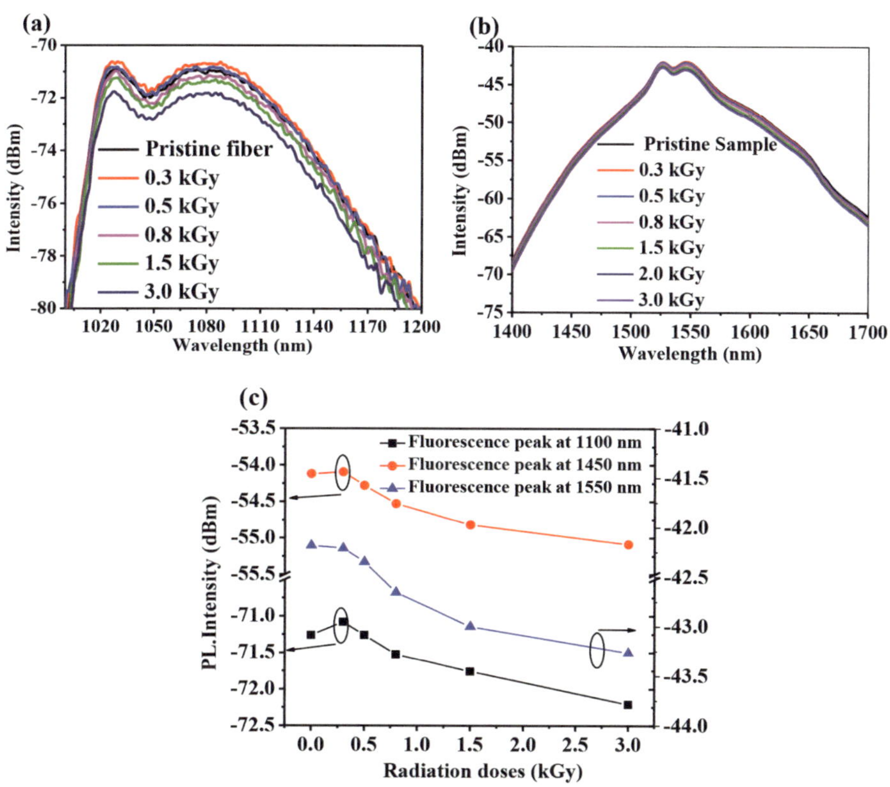

Figure 4.
Fluorescence spectra of BEDF samples at different bands with different radiation doses. (a) 1100, (b) 1550 nm, and (c) the variations of the fluorescence intensity at 1100, 1450, and 1550 nm.

0.5 kGy, the fluorescence intensity of BAC-Al was significantly lower than that of the pristine fiber. In addition, the fluorescence intensity of BAC-Si in BEDF showed the same trend in **Figure 4(c)** (red curve). The fluorescence of Er ions at 1550 nm was also observed, as shown in **Figure 4(b)**. For Er ions, the fluorescence intensity decreased with an increase in the radiation dose and fluorescence enhancement at low-dose radiation (<0.5 kGy) such as Bi ions did not appear..

For BPDF, five BPDF samples were irradiated with cumulative doses of approximately 0.3, 1.0, 1.5, 2.0, and 3.0 kGy at room temperature. The radiation dose rate was 800 Gy/h, which is the same as in the other experiment. The fluorescence spectra of BPDFs at different doses under 830 nm pumping are shown in **Figure 5(a)**. Comparing the PL spectra before and after irradiation, the shape did not change significantly. The fluorescence spectra of the fiber samples range from 1100 to 1600 nm with a peak at 1420 nm, which is derived from BAC-Si. The change in the fluorescence peak of BAC-Si is shown in **Figure 5(b)**. With an increase in the radiation dose, the fluorescence intensities of BAC-Si first increased and then decreased with a further increase in the radiation dose. Moreover, when the radiation dose was 1.5 kGy, the fluorescence intensity of BAC-Si was two times that of the unirradiated BPDF. That is to say, low-dose irradiation can promote the formation of BAC-Si, enhancing the fluorescence intensity. For radiation doses up to 3.0 kGy, the fluorescence intensity of BAC-Si was still higher than that of untreated fiber. This indicated that the BPDF samples had a certain degree of radiation resistance, which has great potential for photonic applications of optical fiber amplification devices in harsh radiation environments.

3.2 Radiation effect on fluorescence lifetime

The luminescence decay curves of the Bi-related active centers in BEDFs and BPDFs were measured using a fluorescence spectrophotometer (Edinburgh FLS-980, England) equipped with an nF900 flash lamp. The fluorescence lifetime decay curves of BAC-Al in BEDF samples before and after radiation are shown in **Figure 6(a)**. In order to compare the fluorescence decay curves of the BEDF samples with different radiation doses, a single exponential function was used to fit them. The relationship between fluorescence lifetime and radiation dose is shown in **Figure 6(b)**. When the radiation doses were 0, 0.3, 0.5, 0.8, 1.5, and 3 kGy, the fluorescence lifetimes of BAC-Al were 564, 599, 585, 560, 559, and 553 μs, respectively. These results demonstrated that their lifetimes increased at low radiation doses (0–0.3 kGy) that were increasing, whereas at higher radiation doses (0.5–3 kGy), their lifetimes were decreased.

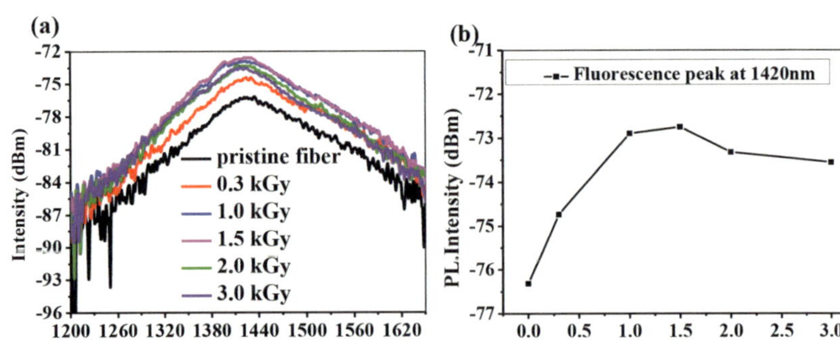

Figure 5.
(a) PL spectra of BPDF samples with different radiation doses and (b) variation of the fluorescence intensity at 1420 nm.

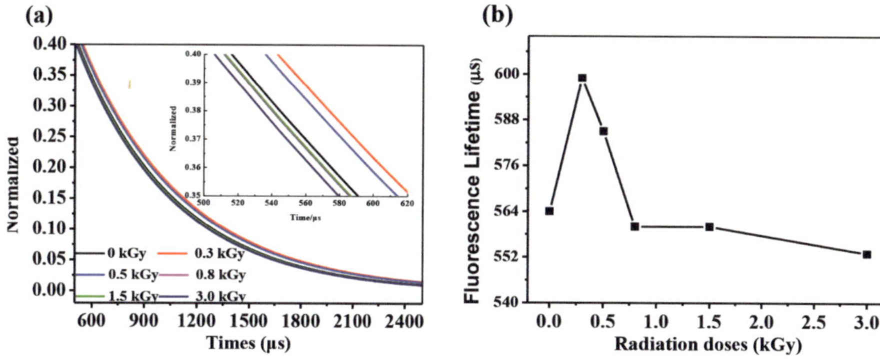

Figure 6.
(a) Luminescence decay curves with different radiation doses and (b) variation in the fluorescence lifetime.

For comparative analysis, the fluorescence lifetime of the Er^{3+} ions at 1534 nm was also measured, as shown in **Figure 7(a)**; when the radiation doses were 0, 0.3, 0.5, 0.8, 1.5, and 3 kGy, the fluorescence lifetimes of the Er^{3+} ions were 11.26, 11.13, 11.11, 11.10, 10.73, and 10.23 ms, respectively. The fluorescence lifetimes of Er^{3+} ions decreased with increasing of radiation doses, as shown in **Figure 7(b)**.

For the BPDF samples, the luminescence decay curves of BAC-Al are presented in **Figure 8(a)**. The single exponential function is a close fit. The luminescence lifetimes of BAC-Al were 740, 699, 573, and 500 μs for radiation doses of 0, 0.3, 1.0, and 3.0 kGy, respectively. Further, under the radiation conditions, the lifetimes of BAC-Al decreased rapidly, as shown in **Figure 8(b)**. It is inferred that the radiation increases the probability of the non-radiative transition, which may be attributed to the faster process whereby the electron in the excited state returns to the ground state or to the role of lead ions. To confirm this hypothesis, a more detailed experiment is required in the future.

3.3 Effect of radiation on unsaturable absorption characteristics

Unsaturable pump absorption (α_{us}) is ideally determined by the direct measurement of the remaining absorption of pump light. The saturable pump absorption (α_s), which is a measure of the effective pump absorption of the fiber used for the radiative emission, decreases with the increasing pump power. The pump absorption consists of α_{us} and α_s. In fact, we focus more on the merit M_α, defined as

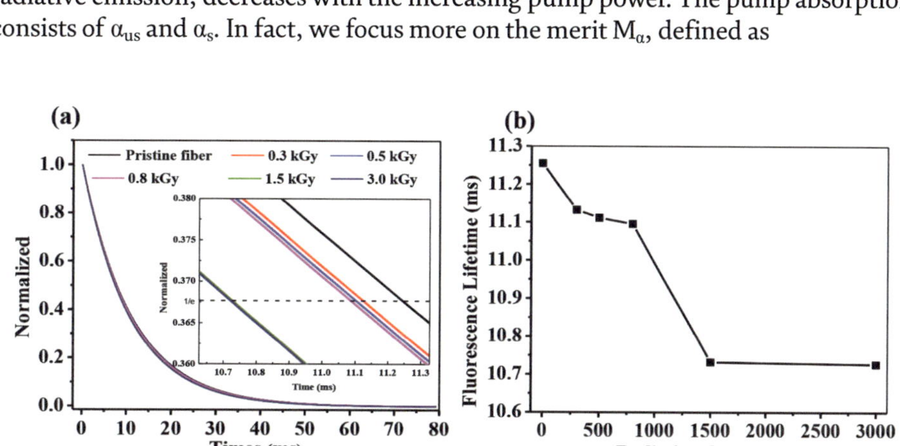

Figure 7.
(a) Luminescence decay curves of Er^{3+} active center in BEDF with different radiation doses and (b) variation of the fluorescence lifetime.

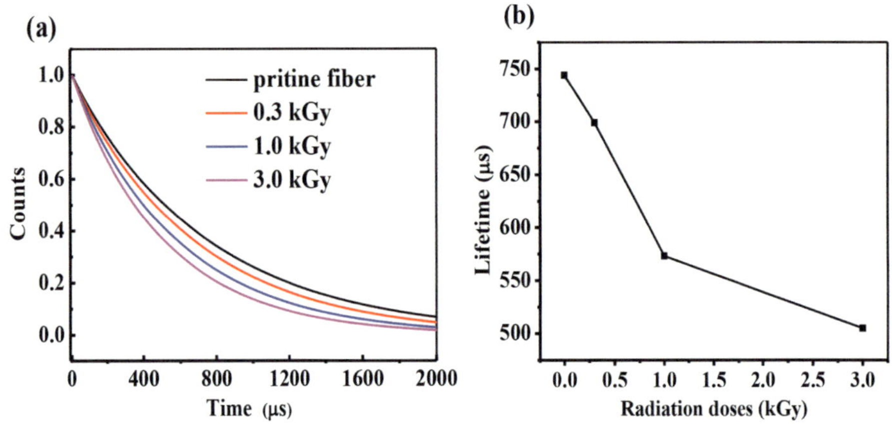

Figure 8.
(a) Luminescence decay curves of BAC-Al in BPDF samples with different radiation doses and (b) variation of the fluorescence lifetime.

$M_\alpha = \alpha_s / (\alpha_s + \alpha_{us})$, which represents the ratio of useful pump absorption, α_s, to the total pump absorption at the pump wavelength. This fraction is a key indicator of useful pump absorption and has a direct correlation to laser efficiency. Here, the unsaturable absorption characteristics of BEDFs at 980 nm before and after irradiation were investigated, as shown in **Figure 9(a)**. When the radiation doses were 0, 0.3, 0.5, 0.8, 1.5, and 3 kGy, the α_{us} values of the BEDF were 40.6, 37.0, 40.7, 43.5, 46.8, and 49.6 dB/m, respectively. As the radiation dose increased, α_{us} first decreased and then increased, as shown in **Figure 9(b)**. According to the relationship between $\alpha_{\mu s}$ and the radiation dose, the decrease of $\alpha_{\mu s}$ in the sample at a low radiation dose (0.3 kGy) may be attributed to the local structural change of Bi ions. Moreover, when the radiation dose was below 3.0 kGy, the α_s of the BEDF (3.6 dB/m) was smaller than that of the unirradiated BEDF. At the same time, their corresponding M_α values were also calculated as 58.6%, 57.5%, 54.8%, 53.3%, 52.2%, and 50.2%. Hence, the M_α of BEDF continuously decreased with an increase in the radiation dose.

The unsaturable absorption characteristics of the BPDF and Bi-doped silica fibers are shown in **Figure 10**. The unsaturable absorption of the Bi-doped silica fiber (α_{us1}) and the Pb/Bi co-doped silica fiber (α_{us2}) at 830 nm were approximately 18 and 8 dB/m, respectively, and their corresponding saturable absorptions were 72 dB/m (α_{s1}) and 45 dB/m (α_{s2}), respectively.

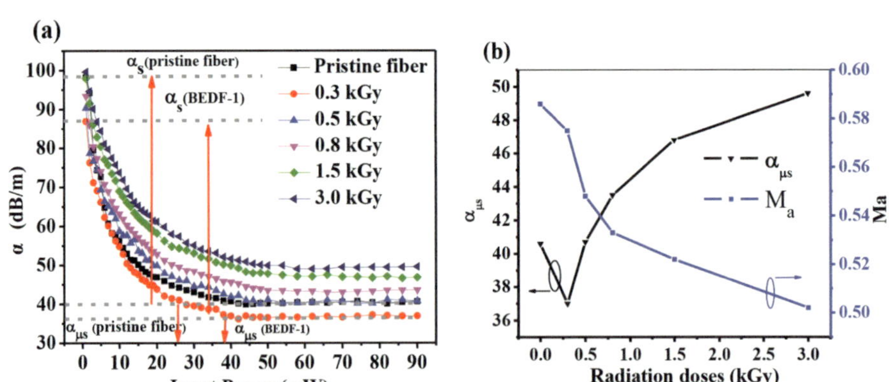

Figure 9.
(a) Unsaturable absorption characteristic of BEDF samples with different radiation doses and (b) variation of the α_{us} *and* M_α.

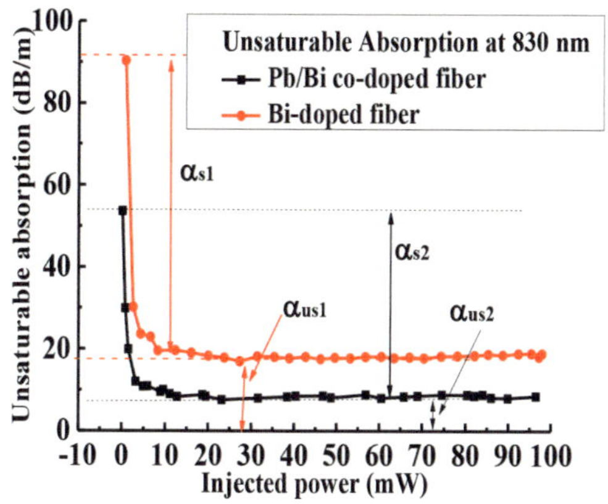

Figure 10.
Unsaturable absorption characteristic of Pb/Bi co-doped fiber (black curve) and Bi-doped fiber (red curve) at 830 nm.

The derived merit M_α of the Pb/Bi co-doped silica fiber was approximately 85.1%, which was larger than that of the Bi-doped silica fiber (80.0%). A high merit M_α meant that a large proportion of the pump photons would participate in the excitation of the active ions, promoting the desirable luminescence process at the corresponding bands. As such, the larger the M_α value, the higher the laser efficiency. Compared with the fiber-doped Bi ions only, the Pb/Bi co-doped silica fiber exhibited improved unsaturable characteristics. This would be beneficial for fiber lasers and amplifiers.

After the BPDF samples were treated with different radiation doses, the unsaturable absorption characteristics were measured as shown in **Figure 11(a)**, and both α_{us} and α_s changed significantly. With an increase in radiation dose, α_{us} trended with a gradual increase, whereas α_s decreased and exhibited a small fluctuation. Furthermore, M_a trended similar to α_s, as shown in **Figure 11(b)**. For the BPDF, the radiation effect on α_{us} was small, similar to the effect of radiation on the fluorescence lifetime of the Er^{3+} ions.

3.4 Effect of radiation on magnetic-optical property

To further study the influence of radiation on the characteristics of Bi ions, the effect of radiation on the magnetic-optical properties of the Bi-doped silica fiber (BDF) was investigated by comparing it with other silica fibers, such as SMF and Pb-doped silica fiber.

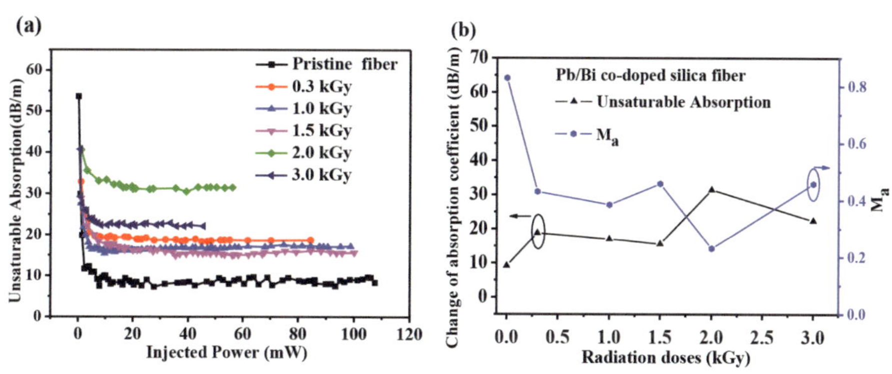

Figure 11.
(a) Unsaturable absorption characteristics of BPDF at 830 nm with different radiation doses and (b) unsaturated absorption coefficient and M_α value with the function of radiation dose.

The Faraday rotation degree of the BDF in different magnetic fields ranging from 0 to 118 mT was measured. The slope of the Faraday rotation curve, marked as β_i, where i = 1–7, in **Figure 12(a)**, determined the Verdet constants of the corresponding fiber samples. The Faraday rotations of the fiber samples were proportional to the intensity of the applied magnetic field. The slope of the rotation angle of BDF (β_2) before irradiation was larger than that of SMF (β_1). After the irradiation, the trend of the slope of the rotation angle changed from β_2 to β_4 clockwise, and then from β_5 to β_7 anticlockwise. The Verdet constant (1.64 rad/(Tm)) of the BDF before irradiation is 26.0% larger than that of SMF (1.29 rad/(Tm)), and the Verdet constant value is positive, indicating that the BDF material has diamagnetic properties. After radiation, the Verdet constant of the SMF increased with increasing radiation doses, as shown in **Figure 12(b)**; however, those of the BDF decreased at low radiation doses (<0.3 kGy). In particular, after 0.3 kGy of irradiation, the Verdet constant of the BDF became negative, showing that the BDF material has a paramagnetic property. Its Verdet constant value was positive and increased with the increase in radiation doses from 0.5 to 3 kGy. The Verdet constant of the BDF after 3.0 kGy of irradiation became 1.87 rad/(Tm), which is 23.84% larger than that of SMF with 1.51 rad/(Tm) and 44.96% larger than that of SMF without radiation.

For the irradiated SMF and Pb-doped silica fibers, their Verdet constants always increased with an increase in the radiation dose, as shown by the red and black curves in **Figure 13**. With a further increase in radiation doses, the Verdet constant of the SMF became essentially constant, which may be due to the fact that the concentration of Ge-related defect centers induced by radiation tended to be saturated. For the Pb-doped silica fiber, the Verdet constant also increased with an increase in the radiation dose (0–1.5 kGy). The Verdet constant of the Pb-doped silica fiber was higher than that of the SMF. This result indicated that gamma-ray radiation enhanced the Verdet constants of the fiber samples, especially for Pb-doped silica fibers. Irradiation not only induced Ge- and Si-related defect centers such as Si′, Ge′ color centers, but also led to new Pb-related defect centers in the Pb-doped silica fibers. These defect centers increased the electron transition probability of Pb^{2+} in $^1S_0 \rightarrow {}^1P_1$ and contributed further to the orbital electron spin. This may be why the increase of the Verdet constant for Pb-doped silica fiber is faster than that for the SMF with an increase in the radiation dose (1.5–2.5 kGy). Therefore, it is supposed that gamma rays improve the magneto-optical properties of fibers.

For the BDF irradiation, with the increase in the radiation dose, the Verdet constant of the BDF decreased first and then increased. In particular, under

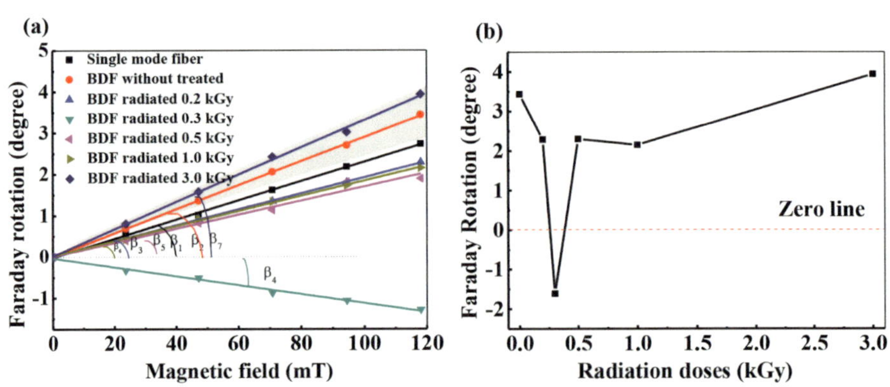

Figure 12.
Relationship between Faraday rotation and (a) magnetic field density and (b) radiation doses.

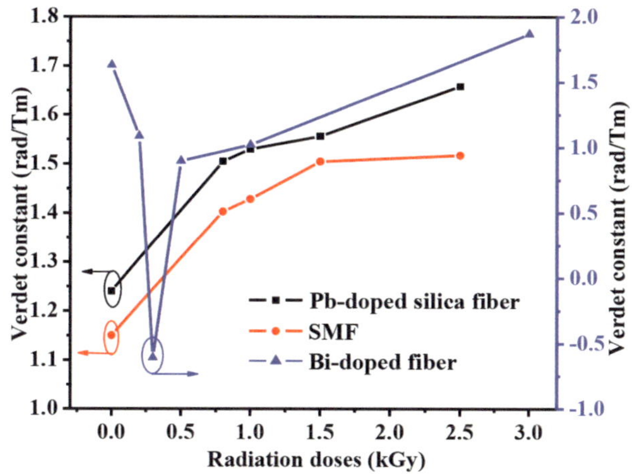

Figure 13.
Verdet constants of Bi-doped silica fiber, Pb-doped silica fiber, and SMF with different radiation doses.

0.3 kGy, the Verdet constant had a negative value, as shown by the blue curve in **Figure 11**. The change in the Verdet constant may mainly result from Bi ions, which present the formation of multiple valence states in the fiber, such as Bi^{0}, Bi^{1+}, Bi^{2+}, Bi^{3+}, and Bi^{5+}. Furthermore, among various valence states, the conversion may be possible under radiation treatment. These different valence states have different outer electronic shell structures. Bi^{3+} ($6s^2$) and Bi^{5+} ($5d^{10}$) ions, which have no unpaired electrons in their outer electronic shells, showed diamagnetic properties. In contrast, Bi^{0} ($6s^2 6p^3$), Bi^{+} ($6s^2 6p^2$), and Bi^{2+} ($6s^2 6p^1$) showed paramagnetic properties because of unpaired electrons in the 6p layer, contributing to the intrinsic magnetic moment. These detailed results have already been reported in [22, 37]. Furthermore, the Verdet constant increase of the Bi-doped silica fiber was faster than that of the SMF and Pb-doped silica fiber with the increase in the radiation dose (1.5–2.5 kGy). Therefore, it is believed that gamma rays clearly improve the magneto-optical properties of the BDF.

4. Conclusion

In this chapter, certain types of BRDFs, including Bi/Al, Bi/Pb, and Bi/Er co-doped optical fibers, were fabricated using the ALD and MCVD process. Then, the radiation effects on their optical properties were investigated. The fluorescence intensity and fluorescence lifetimes of the BRDFs at 1150 nm with low-dose radiation increased significantly, whereas they decreased with a further increase in the radiation dose. The merit M_{α} values of the BRDFs, a ratio of useful pump absorption to total pump absorption, decreased with an increase in the radiation doses. However, the Verdet constants in different doped fibers increased and reached saturation with the increasing radiation dose. The incremental increases of the Verdet constants for the Pb-doped and Bi-doped fibers were faster than those for the SMF with an increase in the radiation dose (1.5–2.5 kGy). Moreover, the Verdet constant decreased and the direction of Faraday's rotation changed at low radiation doses. Hence, the increase in the Verdet constant increase for BDF is much faster than that of other fiber samples treated with high-dose radiation. All these results are of great significance for the study of the optical properties of BRDFs.

Funding

This work is supported by Natural Science Foundation of China (Grant Nos. 61520106014, 61975113, 61935002, and 61675125) and the Pre-Research Fund Project (6140414030203).

Author details

Jianxiang Wen[1*], Ying Wan[1], Yanhua Dong[1], Yi Huang[1], Yanhua Luo[2], Gang-Ding Peng[2], Fufei Pang[1] and Tingyun Wang[1]

1 Key Lab of Specialty Fiber Optics and Optical Access Networks, School of Communication and Information Engineering, Shanghai Institute for Advanced Communication and Data Science, Shanghai University, Shanghai, China

2 Photonics and Optical Communications, School of Electrical Engineering and Telecommunications, University of New South Wales, Sydney, NSW, Australia

*Address all correspondence to: wenjx@shu.edu.cn

References

[1] Fujimoto Y, Nakatsuka M. Infrared luminescence from bismuth-doped silica glass. Japanese Journal of Applied Physics. 2001;**40**(3B):L279-L281

[2] Fujimoto Y. Local structure of the infrared bismuth luminescent center in bismuth-doped silica glass. Journal of the American Ceramic Society. 2010;**93**(2):581-589

[3] Firstov S, Alyshev S, Melkumov M, R iumkin K, Shubin A, Dianov E. Bismuth-doped optical fibers and fiber lasers for a spectral region of 1600-1800 nm. Optics Letters. 2014;**39**(24):6927-6930

[4] Dianov EM. Amplification in extended transmission bands using bismuth-doped optical fibers. Journal of Lightwave Technology. 2013;**31**(4):681-688

[5] Razdobreev I, Bigot L, Pureur V, Favre A, Bouwmans G, Douay M. Efficient all-fiber bismuth-doped laser. Applied Physics Letters. 2007;**90**(3):031103

[6] Bufetov IA, Dianov EM. Bi-doped fiber lasers. Laser Physics Letters. 2009;**6**(7):487-504

[7] Zheng JY, Peng MY, Kang FW, Cao RP, Ma ZJ, Dong GP, et al. Broadband NIR luminescence from a new bismuth doped $Ba_2B_5O_9Cl$ crystal: evidence for the Bi^0 model. Optics Express. 2012;**20**(20):22569-22578

[8] Zhang LL, Dong GP, Wu JD, Peng MY, Qiu JR. Excitation wavelength-dependent near-infrared luminescence from Bi-doped silica glass. Journal of Alloys and Compounds. 2012;**531**:10-13

[9] Sokolov VO, Plotnichenko VG, Dianov EM. Origin of broadband near-infrared luminescence in bismuth-doped glasses. Optics Letters. 2008;**33**(13):1488-1490

[10] Ren JJ, Qiu JR, Chen DP, Wang C, Jiang XW, Zhu CS. Infrared luminescence properties of bismuth-doped barium silicate glasses. Journal of Materials Research. 2007;**22**(7):1954-1958

[11] Xia H, Wang X. Near infrared broadband emission from Bi^{5+}-doped Al_2O_3-GeO_2-X (X=Na_2O,BaO,Y_2O_3) glasses. Applied Physics Letters. 2006;**89**:051917

[12] Ban C, Bulatov LI, Dvoyrin VV, Mashinsky VM, Limberger HG, Dianov EM. Infrared luminescence enhancement by UV-irradiation of H_2-loaded Bi-Al-doped fiber. In: European Conference on Optical Communication, 35th ECOC. Vienna, Austria: IEEE; 2009. p. 2

[13] Violakis G, Limberger HG, Mashinsky VM, Dianov EM. Dose dependence of luminescence increase in H_2-loaded Bi-Al co-doped optical fibers by cw 244-nm and pulsed 193-nm laser irradiation. In: Optical Fiber Communication Conference, OFC. California, United States: IEEE; 2013. P. OTh4C.2

[14] Shen W, Ren J, Baccaro S, Cemmi A, Chen GR. Broadband infrared luminescence in γ-ray irradiated bismuth borosilicate glasses. Optics Letters. 2013;**38**(4):516-518

[15] Firstov S, Alyshev S, Khopin V, Melkumov M, Guryanov A, Dianov E. Photobleaching effect in bismuth-doped germanosilicate fibers. Optics Express. 2015;**23**(15):19226-19233

[16] Ou Y, Baccaro S, Zhang Y, Yang Y, Chen G. Effect of gamma-ray irradiation on the optical properties of PbO-B_2O_3-SiO_2 and Bi_2O_3-B_2O_3-SiO_2 glasses. Journal

of the American Ceramic Society. 2010;**93**(2):338-341

[17] Elbatal FH, Marzouk MA, Abdel-Ghany AM. Gamma rays interaction with bismuth borate glasses doped by transition metal ions. Journal of Materials Science. 2011;**46**(15):5140-5152

[18] Girard S, Kuhnhenn J, Gusarov A, Brichard B, Van Uffelen M, Ouerdane Y, et al. Radiation effects on silica-based optical fibers: Recent advances and future challenges. IEEE Transactions on Nuclear Science. 2013;**60**(3):2015-2036

[19] Wen JX, Liu WJ, Dong YH, Luo YH, Peng GD, Chen N, et al. Radiation-induced photoluminescence enhancement of Bi/Al-codoped silica optical fibers via atomic layer deposition. Optics Express. 2015;**23**:29004-29013

[20] Guo Q , Wen JX, Huang Y, Wang WN, Pang FF, Chen ZY, et al. Magneto-optical properties and measurement of the novel doping silica optical fibers. Measurement. 2018;**127**:63-67

[21] Kim Y, Ju S, Jeong S, Jang MJ, Kim JY, Lee NH, et al. Influence of gamma-ray irradiation on Faraday effect of Cu-doped germano-silicate optical fiber. Nuclear Instruments and Methods in Physics Research Section B: Beam Interactions with Materials and Atoms. 2015;**344**:39-43

[22] Wen JX, Wang WN, Guo Q , Huang Y, Dong YH, Pang FF, et al. Gamma-ray radiation on magneto-optical property of Pb-doped silica fiber. Inorganic Materials. 2018;**33**(4):416-420

[23] Wei S, Luo YH, Ding MJ, Cai FF, Xiao G, Fan DS, et al. Thermal effect on attenuation and luminescence of Bi/Er Co-doped fiber. IEEE Photonics Technology Letters. 2017;**29**(1):43-46

[24] Yang G, Chen DP, Wang W, Xu YS, Zeng HD, Yang YX, et al. Effects of thermal treatment on broadband near-infrared emission from Bi-doped chalcohalide glasses. Journal of the European Ceramic Society. 2008;**28**:3189-3191

[25] Yang G, Chen DP, Ren J, Xu YS, Zeng HD, Yang YX, et al. Effects of melting temperature on the broadband infrared luminescence of Bi-doped and Bi/Dy co-doped chalcohalide glasses. Journal of the American Ceramic Society. 2007;**90**(11):3670-3672

[26] Dianov EM, Dvoyrin VV, Mashinsky VM, Umnikov AA, Yashkov MV, Gur'yanov AN. CW bismuth fibre laser. Quantum Electronics. 2005;**35**:1083-1084

[27] Puurunen RL. Surface chemistry of atomic layer deposition: A case study for the trimethylaluminum/water process. Applied Physics Reviews. 2005;**97**:121301

[28] George SM. Atomic layer deposition: An overview. Chemical Reviews. 2010;**110**:111-131

[29] Sneck S, Soininen P, Putkonen M, Norin L. A new way to utilize atomic layer deposition-case study: Optical fiber manufacturing. In: Proceedings of AVS - 6th International Conference on Atomic Layer Deposition. 2006

[30] Montiel i Ponsoda JJ, Norin L, Ye C, Bosund M, Söderlund MJ, Tervonen A, et al. Ytterbium-doped fibers fabricated with atomic layer deposition method. Optics Express. 2012;**20**:25085-25095

[31] Dong YH, Wen JX, Pang FF, Chen ZY, Wang J, Luo YH, et al. Optical properties of PbS-doped silica optical fiber materials based on atomic layer deposition. Applied Surface Science. 2014;**320**:372-378

[32] Shang YN, Wen JX, Dong YH, Zhan HH, Luo YH, Peng GD, et al.

Luminescence properties of PbS quantum-dot-doped silica optical fibre produced via atomic layer deposition. Journal of Luminescence. 2017;**187**:201-204

[33] Dong YH, Wen JX, Guo Q, Pang FF, Wang TY. Formation and photoluminescence property of PbS quantum dots in silica optical fiber based on atomic layer deposition. Optical Materials Express. 2015;**5**(4):712-719

[34] Wang Q, Wen JX, Luo YH, Peng G-D, Pang FF, Chen ZY, et al. Enhancement of lifetime in Er-doped silica optical fiber by doping Yb ions via atomic layer deposition. Optical Materials Express. 2020;**10**:2397-2407

[35] Shen YD, Li YW, Li WM, Zhang JZ, Hu ZG, Chu JH. Growth of Bi_2O_3 ultrathin films by atomic layer deposition. Journal of Physical Chemistry C. 2012;**116**:3449-3456

[36] Wang TY, Wen JX, Luo WY, Xiao ZY, Chen ZY. Influences of irradiation on network microstructure of low water peak optical fiber material. Journal of Non-Crystalline Solids. 2010;**356**:1332-1336

[37] Wen JX, Che QQ, Dong YH, Guo Q, Pang FF, Chen ZY, et al. Irradiation effect on the magneto-optical properties of Bi-doped silica optical fiber based on valence state change. Optical Materials Express. 2020;**10**(1):88-98

Chapter 5

Effects of Electron Irradiation on Optical Properties of Bismuth-Doped Phosphosilicate Fiber

Alexander V. Kir'yanov and Arindam Halder

Abstract

The basic optical properties of yttrium-phosphosilicate fiber doped with bismuth (Bi) are assessed in both pristine state and that established after bombardment by a beam of high-energy electrons. The fiber has been developed and fabricated with a target to use it for laser applications in visible/near-infrared (VIS/NIR) domain. In this chapter, the main attention is paid to the dramatic changes in absorption spectra of the fiber under electron irradiation. Meanwhile, we reveal its overall resistance to irradiation in terms of emissive potential and bleaching contrast at excitation into the absorption bands of bismuth-related active centers. Besides, we report a new effect of large dose-dependent Stokes shift, experienced by the fiber's cutoff wavelength, which arises due to refractive index rise in its core area. The laws obeyed by the fiber's characteristics vs. dose are examined for possible applications in dosimetry.

Keywords: bismuth, yttrium-phosphosilicate fiber, electron irradiation, optical properties, cutoff wavelength

1. Introduction

Silica-based fibers doped with semi-metals, e.g., bismuth (Bi), have become a popular object for sensing ionizing radiation (dosimetry), including electron beams: recently some of the Bi-doped fibers (BDFs) were primarily examined in this regard [1–8]. Sensing dose of ionizing radiation using optical fiber is an important issue: dose-induced changes in its characteristics are worth assessing as they may serve a base for making cost-effective, remotely monitorable sensors, easily located in harsh environments such as proximity to a nuclear reactor [9]; optical fibers can be also used for plasma diagnostics in fusion reactors [10]. The relevance of choosing BDFs for sensing relies on the understanding that Bi is highly susceptible to interaction with nuclear particles, e.g., energetic (further "β") electrons. Since high-energy "primary" β-electrons are virtually non-dissipating at propagation through silica fiber, the basic effect behind the transformations arising in its material and optical properties is creating β-induced carriers, viz., "secondary" free holes/electrons. Given that Bi is a heavy semi-metal, it becomes—when used as dopant—a powerful source of secondary carriers, eventually stabilized in core glass network generating plenty of defect centers. Such centers (a part of which is associated with the presence of Bi ions) are responsible for the increasing dose-dependent-induced absorption (IA) in BDFs and, accordingly, modifications of absorption and emission bands of Bi-related active centers (BACs). So far, BDFs with core glass of various chemical compositions were examined in this regard. Particularly, it was demonstrated that BDFs with aluminosilicate and germanosilicate hosts are highly susceptible to different types of ionizing radiations [1–8], with the result being notable growth of IA in visible/near-infrared (VIS/NIR) and severe changes in the absorption bands of the respective BACs. Say, it was found that Bi/Ge and Bi/Si BACs fade while Bi/Al BACs are born posterior to β-irradiation. However, the case of Bi-doped phosphosilicate fiber remained untouched.

In this chapter, we present a review of our recent studies of the basic material and optical properties of a homemade Bi-doped yttrium-phosphosilicate (BYP) fiber in pristine state [11–13] and after irradiating by a beam of β-electrons [14]. The review's main scopes are dose-induced transformations in absorptive, emissive, and waveguide characteristics of the fiber. The data presented below can serve a base for its use for dosimetry purposes.

2. Optical properties of BYP fiber in pristine state

BYP fiber (its cleaved-end image is shown in **Figure 1**) was drawn from the preform, fabricated by employing modified chemical vapor deposition process along with solution-doping techniques; the details are reported in [11, 12].

The basic characteristics of BYP fiber are summarized in **Table 1**.

The cross-sectional distributions of the doping elements (P, Y, and Bi) within the core region of BYP preform were measured using electron microprobe analysis and are plotted in **Figure 2**; their contents were found to be 5.26, 0.14, and 0.07 wt.%, respectively.

The basic waveguide properties of the fiber, viz., numerical aperture (NA), ~0.11; core diameter, ~4.42 μm; and refractive-index (RI) difference, ~0.004, were managed to fulfill single-mode propagation of light beyond ~600 nm, which is potentially perspective for amplifying/lasing in VIS/NIR [13].

The RI transversal distribution of BYP fiber was obtained using a fiber profilometer and is shown in **Figure 3**.

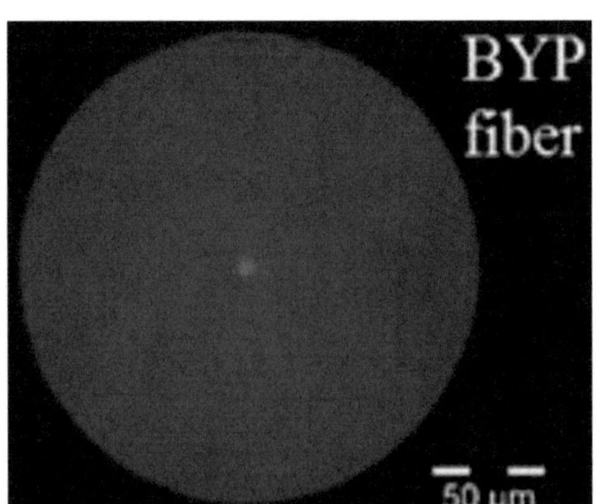

Figure 1.
Cross-sectional image of BYP fiber.

	Dopants' concentrations (wt.%)	Core/cladding diameters (μm)	NA	Cutoff wavelength (nm)
BYP fiber	P = 5.26 Y = 0.14 Bi = 0.07	4.4/123.3	0.12	645

Table 1.
Optical properties of BYP fiber.

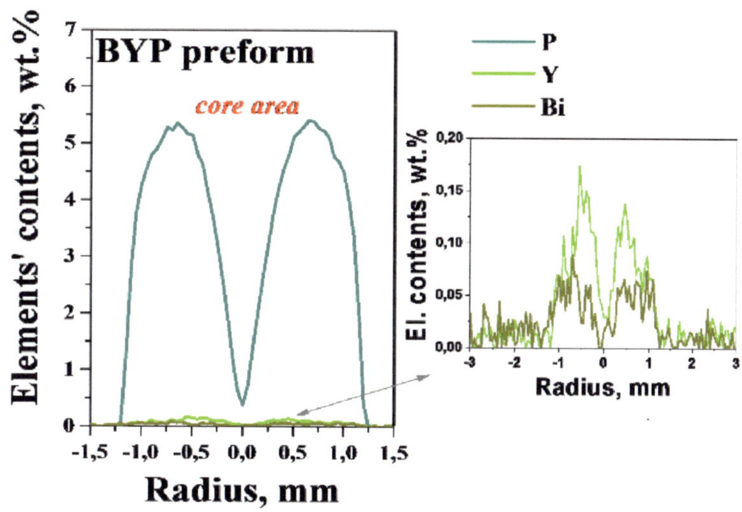

Figure 2.
Cross-sectional distributions of the dopants in BYP fiber preform.

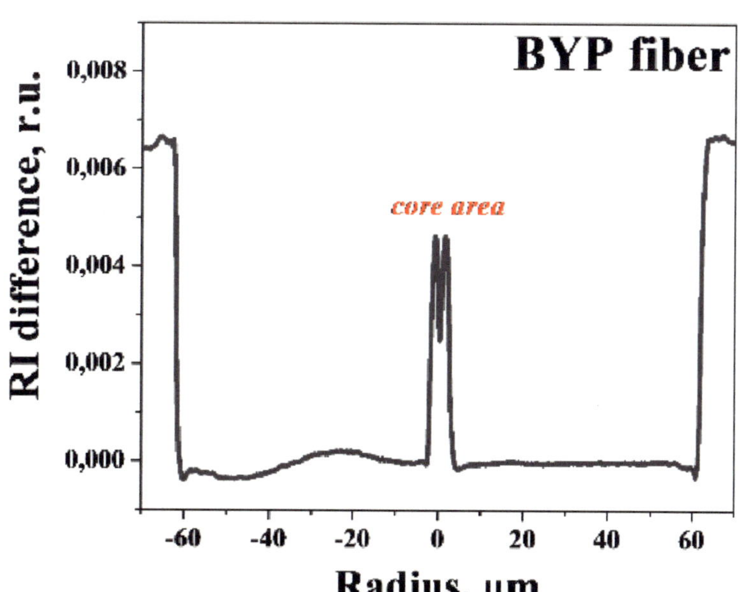

Figure 3.
RI profile of BYP fiber.

The fiber's spectral attenuations in straight and coiled positions were measured, and their difference was found; the result is shown in **Figure 4**. As seen from the figure, the stop wavelength of spectral oscillations is at ~645 nm (in fact, it is the cutoff wavelength λ_c), beyond which the fiber supports single-mode propagation ($LP_{0,1}$ mode).

The absorption, or loss, spectrum $\alpha_0(\lambda)$ of BYP fiber (the solid line in **Figure 5**) was measured using a white-light (WL) source with fiber output and optical spectrum analyzer (OSA) by the cutback method. As seen, the main spectral features of its spectral attenuation (i.e., the broad absorptions bands in VIS/NIR) match the ones known for Bi-related (Bi/P and Bi/Si) BACs. As also seen from the figure, these bands are superimposed with the background (base) loss, steadily growing toward shorter wavelengths, a well-known feature for BDFs. This background loss was recently shown to be explained by Rayleigh scattering (RS) on Bi nanoparticles in the core region [12]; see the dashed line in **Figure 5**.

Photoluminescence (PL) spectra of BYP fiber were recorded in "forward" geometry at in-core pumping using a set of low-power laser diodes (LDs) with fiber outputs. The pump wavelengths explored were 405, 520, 633, 720, 905, and 976 nm (see the arrows in **Figure 5**, the colors of which are congruent with the spectral subdomains where excitation lights fall in). Lengths (L_f) of fiber samples were chosen to provide comparable optical densities (OD = $\alpha_0 L_f$) at these wavelengths, in accord with spectrally variable extinction in the BAC bands. Note that OD was chosen to be ~2.5, providing effective pump absorption and, in the meantime, a little effect of PL reabsorption. The resultant PL spectra are shown in **Figure 6**, all recorded at the same incidence power at each pump wavelength (~3.5 mW). Such pump level was chosen, on the one hand, to provide population inversion in the system of BACs sufficient to get pronounced PL power and, on the other hand, to guarantee safety of registration with OSA. As seen, BYP fiber demonstrates pronounced PL in VIS/NIR, characteristic to BACs in phosphosilicate fibers [1, 11–13],

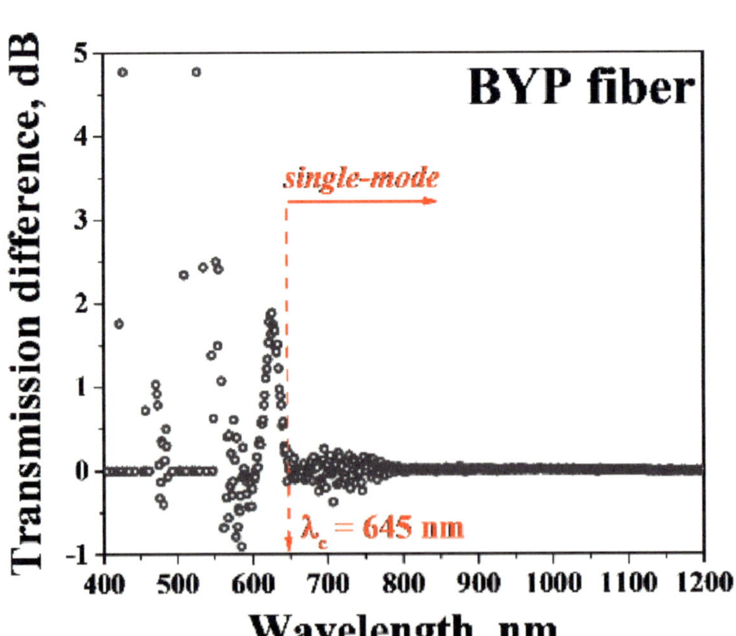

Figure 4.
Transmission difference (straight to coiled positions) of BYP fiber, revealing the spectral position of the cutoff wavelength (~645 nm).

with maximum at ~1200–1300 nm slightly depending on the excitation wavelength. Say, when pumped at 405 nm, BYP fiber demonstrates the PL bands centered at 760 and 1280 nm. In turn, when pumped in NIR (at 730 and 905 nm), it emits the second band only, whose gravity center is barely shifting to shorter wavelengths with increasing pump wavelength.

The behavior of nonlinear, viz., in function of pump power P_{in}, absorption, or loss, of BYP fiber, parametrized for the same set of pump wavelengths, is illustrated in **Figure 7**. Experimentally, pump light from a LD was launched through splice to a fiber sample with length L_f (chosen as reciprocal of α_0; hence, L_f were different for each kind of excitation). Its nonlinear transmission was measured as $T_{NL} = P_{out}/P_{in}$ (where P_{in} and P_{out} are the input and output pump powers, respectively), and nonlinear absorption (loss) was then found as $\alpha_{NL} = -10\log(T_{NL})/L_f$. The dependences α_{NL} *vs*. P_{in}, obtained for different pump wavelengths, are plotted in **Figure 7**.

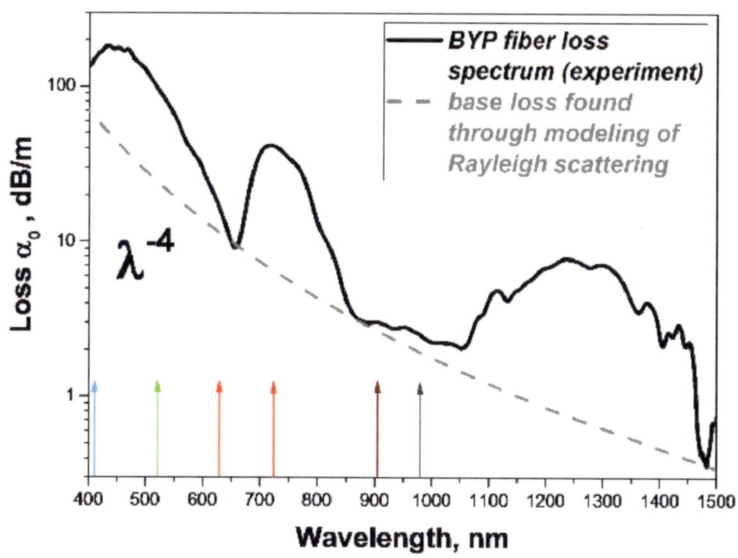

Figure 5.
Experimental absorption spectrum of BYP fiber (solid line) and modeled RS loss spectrum. The excitation wavelengths used in the experiments are marked by arrows of different colors.

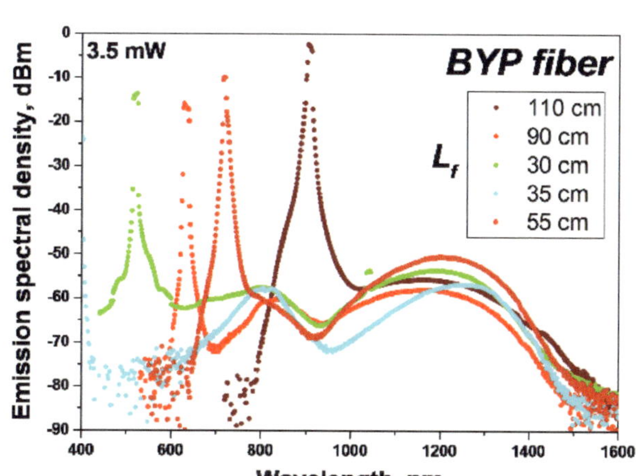

Figure 6.
PL spectra at BYP fiber's output, obtained at different excitation wavelengths (405, 520, 633, 720, and 905 nm); lengths of fiber pieces used are given in the inset.

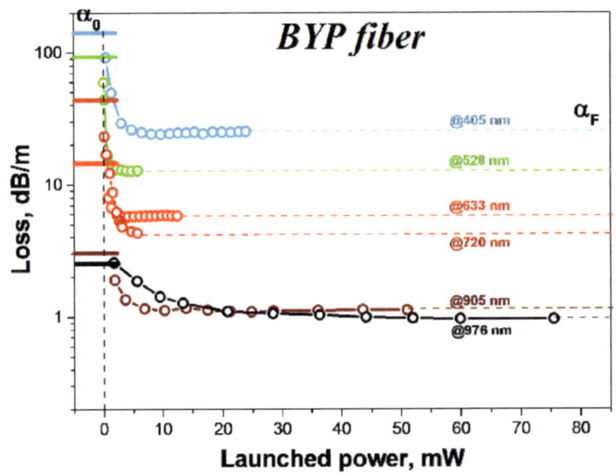

Figure 7.
Dependences of nonlinear loss α_{NL} vs. launched pump power Pin *in BYP fiber, built for all excitation wavelengths (highlighted by relevant colors). The small-signal absorptions* (αo) *at these wavelengths were found from* **Figure 5**. *The residual losses* (αF) *are marked by the dashed lines.*

Note that the plateaus (marked as α_F in the figure), to which the dependences approach at higher pump powers, attribute the unbleached, or residual, absorptions. As seen, for the shorter wavelengths, the loss plateaus (α_F-values) lie above those for the longer ones.

Unbleached loss in the BYP fiber might arise in the system of BACs due to a line of processes such as excited-state absorption (ESA), up-conversion in clustered BACs, scattering on nano-inclusions, etc., thanks to a "cumulative" effect of the mentioned contributions. Generally, it is difficult to undoubtedly segregate the true origin of the effect. Meanwhile, as demonstrated in [12] for the BDFs invented by our team, RS loss may stand behind the phenomenon. For example, for BYP fiber, if the residual losses α_F are plotted in double logarithmic scale vs. excitation wavenumber (reciprocal of wavelength λ: $10^4/\lambda$), the data obey the $\sim\lambda^{-4}$ law with high accuracy. This points on that RS can be a good explanation for growth of the residual (unbleached) loss with decreasing wavelength, $\alpha_F(\lambda)$.

3. Optical properties of BYP fiber after electron irradiation

A controllable linear accelerator emitting β-electrons with a narrow-band energy spectrum (~6 MeV) in a short-pulse (~5-μs) mode was used for irradiating BYP fiber [1]. Its samples ~2 m in length were placed inside the accelerator's chamber for the growing fluence of β-irradiation. Prior to optical measurements, the irradiated samples were relaxed for a month at room temperature to minimize the role of short-decay components in IA; for the same reason, the measurements were done during the following short period of time (2–3 weeks). Five irradiation fluxes (doses) ("1" to "5") were chosen to exemplify the type of changes in optical properties of the fiber, established after β-bombardment: 3×10^{11}, 1×10^{12}, 3×10^{12}, 1×10^{13}, and 1×10^{14} cm^{-2} [69, 230, 6.9×10^2, 2.3×10^3, and 2.3×10^4 Gy(SiO_2)]; its pristine state is referred to as "0" [14].

The basic trends in dose-related modifications of the optical properties of BYP fiber can be captured from **Figures 8–15**.

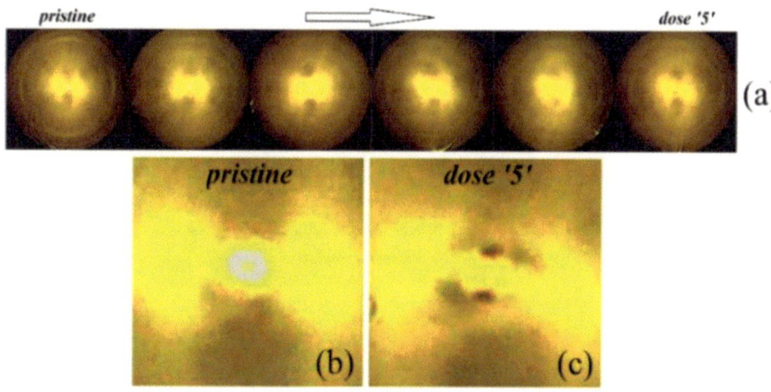

Figure 8.
(a) Cross-sectional images of BYP fiber, obtained for different irradiation doses; the arrow highlights dose increasing. (b and c) insights to core region for BYP fiber in pristine state and after experiencing maximal irradiation dose ("5"), respectively.

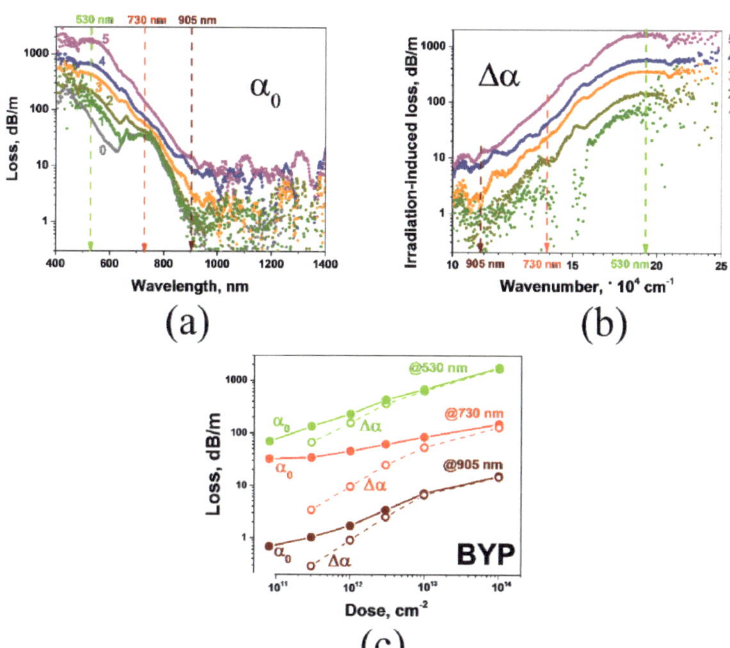

Figure 9.
(a) Attenuation spectra (αo) in VIS/NIR vs. wavelength and (b) IA spectra (Δα) vs. wavenumber for BYP fiber in pristine state (curves "0") and after growing β-irradiation dose (curves "1" to "5"). (c) Dose dependences of small-signal absorption (αo, solid curves) and IA (Δα, dashed curves), measured at wavelengths 905, 730, and 530 nm.

In **Figure 8**, we demonstrate how β-bombardment affects WL transmission of the fiber. It is seen that it suffers darkening (steadily increasing with growing irradiation dose) within the core area but its un-doped silica cladding is barely affected by irradiation.

In **Figure 9**, the absorption spectra vs. wavelength (a) and wavenumber (b) are shown, measured for BYP fiber in pristine state and after β-irradiation with doses 1 to 5 (curves "0" to "5"). As seen, IA magnitudes ($\Delta\alpha$) are very large in the fiber, especially in UV/VIS where IA reaches >2000 dB/m at maximal dose 5.

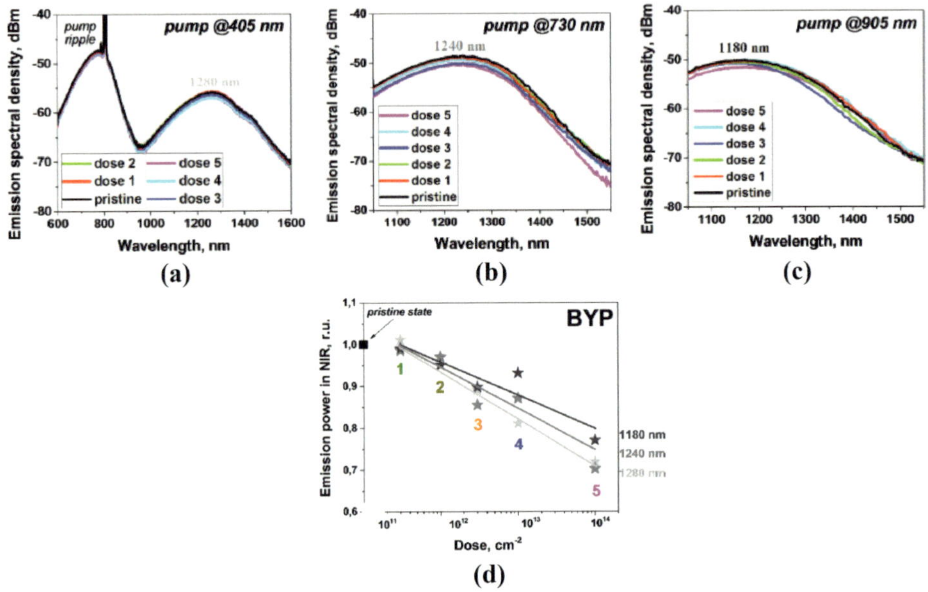

Figure 10.
(a–c) PL spectra of BYP fiber at optical excitation at 405 (a), 730 (b), and 905 (c) nm with launched powers ~25, ~20, and ~100 mW, respectively. (d) Dose dependences of relative (to pristine state) PL powers in the maxima of BAC bands shown in panels (a–c).

It is natural to ascribe the dramatic IA rise in BYP fiber to Bi/P co-doping (doping with Y plays a minor role as "light" Y atom is nearly irresponsible for creating defect centers by nuclear particles.) The spectrally dependent IA in the fiber can be explained likewise that in β-irradiated aluminosilicate BDFs [1, 2], its rise in VIS seemingly originates from extra Bi/P BACs generated, while its gravity center shift toward UV is related to growth of P-centers. Besides, possible contribution in the latter trend may stem from growth of RS (yet occurring in pristine BYP fiber; see [13]) due to irradiation-induced nanoscale degradation of core glass. However, given that IA spectra of irradiated BYP fiber are essentially structureless, it is nearly impossible to separate the evoked mechanisms. The dose-dependent trends for overall loss (α_0) and IA ($\Delta\alpha$), specified for wavelengths 530, 730, and 905 nm [see the dashed vertical lines in **Figure 9(a)** and **(b)**] are resumed in **Figure 9(c)**. Tentatively, the monotonic changes with the dose of both quantities are promising for dosimetry.

In **Figure 10(a)–(c)**, we exemplify the spectral transformations in BAC PL at pumping the fiber at 530, 730, and 905 nm for the growing dose of β-irradiation. In turn, we summarize in **Figure 10(d)** the dose-related changes in PL power. In this case, the same set of equipment and measurement techniques were employed as when studying pristine BYP fiber (refer to **Figure 6** and the text therein). As seen from **Figure 10**, the emissive potential of BYP fiber at either excitation wavelength decreases but barely with dose (not exceeding ~15% at the highest received dose 5, ~10^{14} cm^{-2}). Seemingly, this happens because of generating of extra emission-active Bi/P BACs during irradiation (like it happens with Bi/Al BACs in Bi-doped aluminosilicate fiber [1, 2]), a process leveling the deteriorating effect of IA growth on other defect centers. Thus, BYP fiber is quite resistant to β-irradiation, suggesting its usefulness for remote dosimetry, e.g., at a nuclear plant.

In **Figure 11**, we specify for BYP fiber the bleached part of absorption under pumping at different wavelengths [405 (a), 730 (b), and 905 (c) nm], parametrized

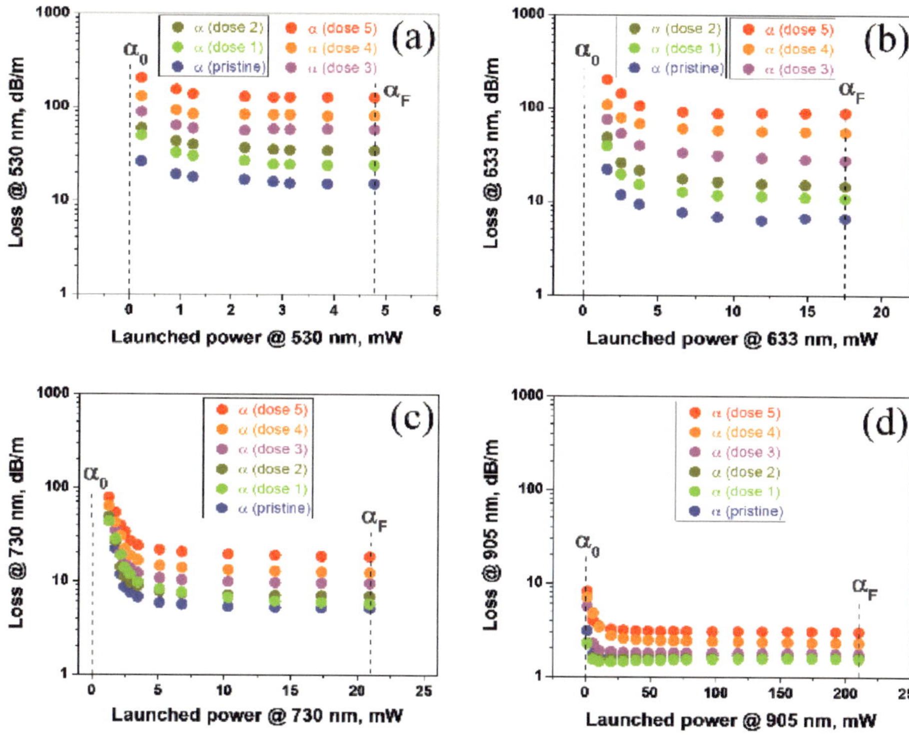

Figure 11.
Nonlinear absorption coefficients of BYP fiber in pristine state and after different doses of β-irradiation (curves 1 to 5) vs. launched LD pump power at 530 (a), 633 (b), 730 (c), and 905 (d) nm.

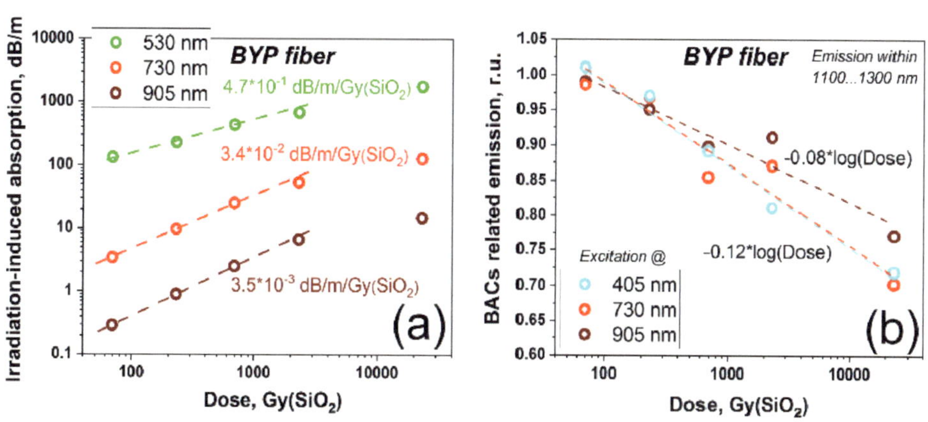

Figure 12.
Prospects for dosimetry using BYP fiber, based on (a) IA rise at 905, 730, and 530 nm and (b) Bi-related emission drop at excitations at 905, 730, and 405 nm vs. β-irradiation dose (symbols and dashed lines correspond to experimental data and their fits, respectively). The dose laws are provided near the experimental dependences: in (a) the slopes of the linearized fits within different spectral domains and in (b) the formula for fitting. All characteristics are given against dose in Gy[SiO_2].

for different irradiation doses (see the symbols of different colors in the panels). The experiments were done by the same way as described above for the pristine BYP fiber (see **Figure 7** and the text therein). As seen from **Figure 11**, all the dependences $\alpha_{NL}(P_{in})$, including the values of small-signal (α_0) and residual (α_F) losses for the wavelengths examined, are all shifted upward with increasing irradiation

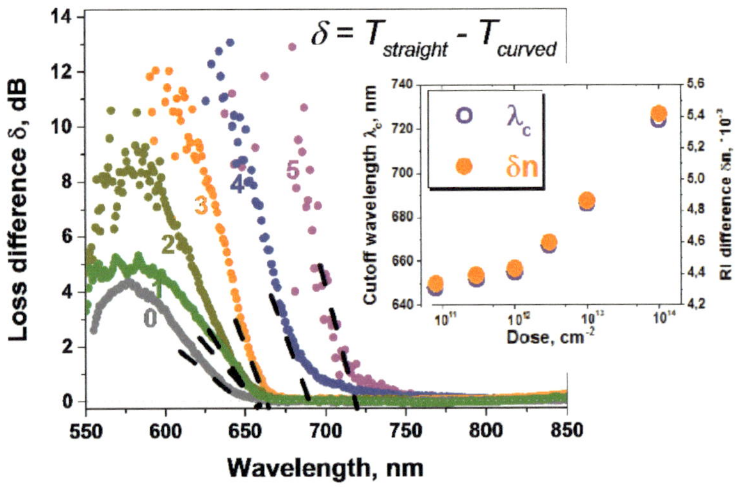

Figure 13.
Main frame: transmission difference spectra of BYP fiber in pristine state and after different irradiation doses, measured at arranging it straight and coiled (for diameter of 5 cm); the cutoff wavelengths λ_c are defined by intersections of the black dashed lines with the horizontal axis. Inset: dose dependences of cutoff wavelength λ_c (left axis, open violet circles, left axis) and RI difference δn (right scale, orange-filled circles).

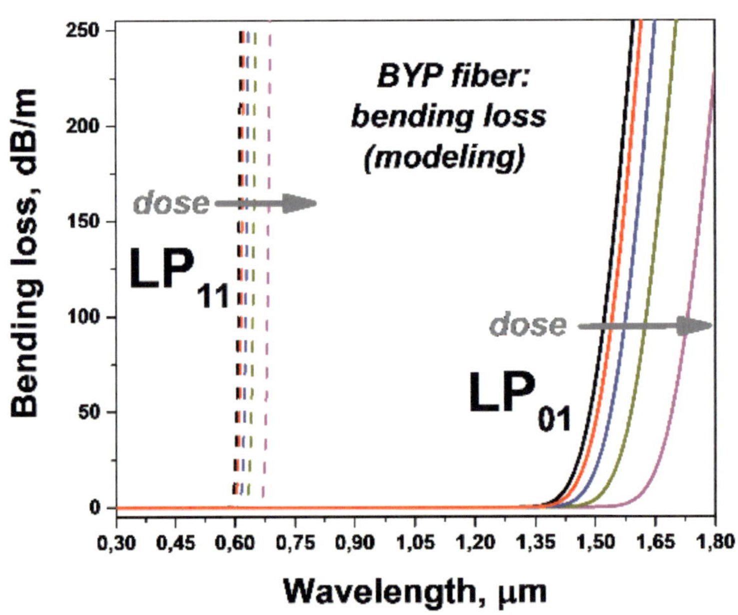

Figure 14.
Results of simulation of bending losses for LP_{01} and LP_{11} modes of BYP fiber bended for 30-cm diameter in pristine state and for the growing doses from 2 to 5 (the arrows highlight dose increasing).

dose; however, the longer the excitation wavelength, the effect is weaker. We suppose that the growth of unbleached loss α_F with dose is explained by boosted scattering in the system of BACs (see above) or/and by P_1-centers produced during irradiation [15, 16]. In the meantime, the growth of bleached part of absorption α_0 with dose arises due to generation of extra Bi/P BACs, which prevails that of residual loss α_F. This trend seems to be advantageous for laser applications of BYP fiber in harsh environments.

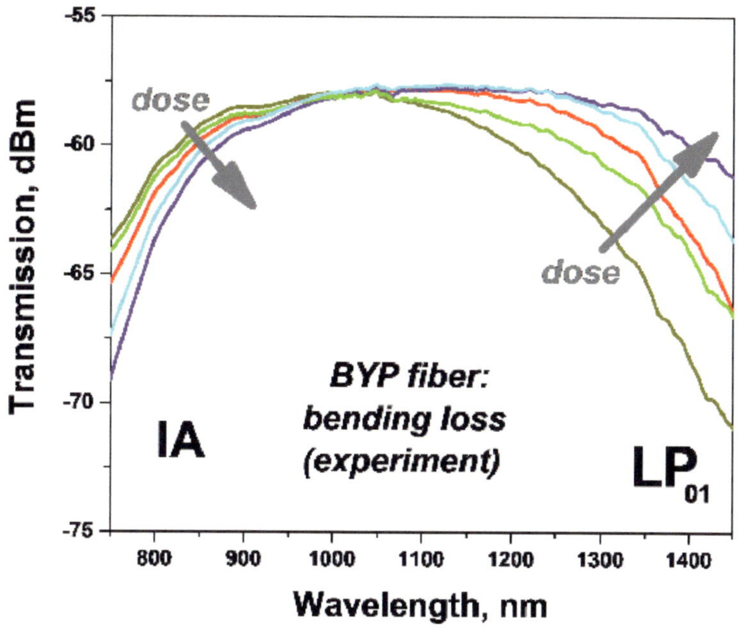

Figure 15.
Experimental transmission spectra of 12-cm BYP fiber samples, similarly bended, being in pristine state and for the growing doses from 2 to 5 (the arrows highlight dose increasing); the spectral ranges where dominate IA and partial escaping to cladding of LP_{01} mode are on the left and right, respectively.

In **Figure 12**, we provide a resume of the basic set of data (reported above) as a base for farther dosimetry assessments of BYP fibers. Note that here the changes in the characteristics under scope are related to irradiation dose counted in Gy[SiO_2] (instead of fluence in cm^{-2}), for comparison of some of them with those previously published for Bi,Y-free phosphosilicate fibers [17, 18].

Regarding the IA growth with dose in BYP fiber, it is seen from (a) that in *log–log* plot this quantity changes with dose up to ~5 kGy[SiO_2] by a nearly linear law but with slope reduced by approximately two orders of value from NIR to VIS. Notably, this trend in NIR range (where BAC absorption is less prominent than in VIS) is less expressed than that reported for un-doped phosphosilicate fiber exposed to X-rays in [17].

Regarding the reduction of NIR PL specific to BACs (within 1.1–1.3 μm) with dose, it is seen from (b) that it steadily diminishes with slope, slightly dependent upon pump wavelength used. The found law inherent to BYP fiber can be used for dosimetry for a wide interval of doses (tens Gy[SiO_2] to tens kGy[SiO_2]).

Another property of BYP fiber, viz., the character of modal wave guidance, was found to severely change with irradiation dose, too. In this regard, a new effect of large Stokes shift, experienced by cutoff wavelength of the fiber vs. dose, was uncovered; its main sides can be revealed from **Figures 13–15**.

Experimentally, we measured bending loss of BYP fiber at each stage of β-irradiation. Probing the fiber with WL, we found spectrally dependent losses in its pristine and irradiated (with different doses) states when it is placed straight and coiled (for ~5-cm diameter); then we found their difference. The resultant spectra are shown in the main frame of **Figure 13**. Note that the last to the Stokes side oscillations in the figure correspond to the first cutoff wavelengths (λ_c) of the fiber after its exposure to different doses (these are marked by the dashed black lines). Beyond these spectral points, only fundamental LP_{01} mode is guided in the fiber. As one can reveal from the figure, λ_c continuously moves to the Stokes side with dose, with maximal shift observed for dose 5 (~75 nm relatively to its position in pristine state of the fiber): refer to the inset in **Figure 13**.

We may infer that the described phenomenon can be eventually explained by dose-dependent rise of BAC-related absorption in core region of BYP fiber, which, in accordance with Kramers-Kronig relations, ought to be accompanied by strong change (growth) of RI. Since darkening of the fiber in core area is measured by hundreds of dB/m around the cutoff wavelength (see **Figure 9**) while un-doped silica cladding is barely affected by irradiation (see **Figure 8**), we consider in the first approximation that core-cladding RI difference vs. dose is entirely defined by the changes arisen in the fiber core. λ_c-values, shown by violet symbols in the inset to **Figure 13**, can be recalculated into RI differences (δn), shown by orange symbols (see the right axis in the same plot). Seemingly, the found effect is pronounceable in BYP fiber due to weak RI difference (~4.3×10^{-3}) in its pristine state.

Let us gain insight into the bending loss issue by modeling of the waveguide properties of BYP fiber. Handling the measured RI profile of BYP fiber [11], we performed, using the software *Optiwave*, an analysis of the modes supported in its core region. An example of the results, obtained for the losses suffered by modes LP_{01} and LP_{11} modes, is demonstrated in **Figure 14**.

It is seen from **Figure 14** that the modeling results for bending loss for LP_{11} mode, viz., both the initial spectral position of cutoff wavelength in pristine BYP fiber and its movement with irradiation dose to longer wavelengths, qualitatively agree with the experiment (compare with **Figure 13**). Besides, the modeling results for bending loss for LP_{01} (fundamental) mode reveal a positive role of β-irradiation in the enhancement of transmittivity of this mode in the NIR. Also, our modeling allows one to understand one of the factors behind these trends, namely, better confinement of both LP_{01} and LP_{11} modes in the core area of the fiber with increasing irradiation dose [14].

To further cross-check the modeled and experimental circumstances of waveguiding and transmittivity in BYP fiber, **Figure 14** can be compared with **Figure 15** showing the transmission spectra of the fiber in pristine and irradiated states, coiled for diameter 30–40 cm. As seen from **Figure 15**, the pattern of WL transmittivity is a superposition of two basic trends, highlighted by the gray arrows: on the one hand, above ~1 μm, it is the rising of LP_{01} mode transmission due to enhanced confinement in the core region (compare with **Figure 14**), and, on the other hand, below ~1 μm, it is the worsening of transmission (merely due to IA).

The results shown above can be adopted, using the principles of modal interference, for making a simple BYP fiber-based sensor of ionizing radiation.

4. Conclusions

The general pattern of the changes in the basic optical characteristics of bismuth-doped yttrium-phosphosilicate fiber as the result of its irradiation by a beam of high-energy electrons is presented. Despite the dramatic rise of induced absorption in the fiber as the result of electron irradiation, we reveal its remarkable resistance, even at the highest doses of exposure, in terms of the emissive potential in VIS/NIR. The found laws of dose-dependent changes in the fiber's optical properties can be useful for dosimetry in harsh conditions, e.g., at a nuclear plant. It is separately emphasized for this type of bismuth-doped fiber the presence of a new effect, consisting in significant Stokes shift, experienced by its cutoff wavelength, proportional to the irradiation dose. As demonstrated, this effect, mainly associated with an increase of the fiber's refractive index as the result of electron irradiation, can be also useful for the dosimetry purposes.

Acknowledgements

The authors are grateful to E. Sekiya and K. Saito (Japan), Y. Barmenkov and V. Minkovich (Mexico), S. Didenko, S. Legotin, and K. Tapero (Russian Federation), and J. Sahu (UK), for their valuable cooperation and helpful discussions. A. Kir'yanov acknowledges the support from the Increase Competitiveness Program of NUST "MISIS" of the Ministry of Education and Science (Russia) (Grant K3-2018-2023).

Author details

Alexander V. Kir'yanov[1,2]* and Arindam Halder[3,4]

1 Centro de Investigaciones en Optica, Leon, Mexico

2 National University of Science and Technology (MISIS), Moscow, Russia

3 Frontier Materials Laboratory, Toyota Technological Institute, Nagoya, Japan

4 Optoelectronics Research Center, University of Southampton, Southampton, UK

*Address all correspondence to: kiryanov@cio.mx

References

[1] Kir'yanov AV. Effects of electron irradiation upon absorptive and fluorescent properties of some doped optical fibers. In: Monteiro WA, editor. Radiation Effects in Materials. InTechOpen; 2016. ISBN 978-953-51-2418-4. DOI: 10.5772/61498

[2] Kir'yanov AV, Dvoyrin VV, Mashinsky VM, Il'ichev NN, Kozlova NS, Dianov EM. Influence of electron irradiation on optical properties of bismuth doped silica fibers. Optics Express. 2011;**19**:6599-6608

[3] Wen J, Liu W, Dong Y, Luo Y, Peng GD, Chen N, et al. Radiation-induced photoluminescence enhancement of Bi/Al-codoped silica optical fibers via atomic layer deposition. Optics Express. 2015;**23**:29004-29013

[4] Firstov SV, Khopin VF, Alyshev SV, Firstova EG, Riumkin KE, Melkumov MA, et al. Effect of gamma-irradiation on the optical properties of bismuth-doped germanosilicate fibers. Optical Materials Express. 2016;**6**:3303-3308

[5] Firstov SV, Khopin VF, Kharakhordin AV, Alyshev SV, Riumkin KE, Melkumov MA, et al. Radiation-induced absorption in bismuth-doped germanosilicate fibres. Quantum Electronics. 2017;**47**:1120-1124

[6] Kir'yanov AV, Barmenkov YO, Minkovich VP, Das S, Dutta D, Dhar A, et al. Effect of electron irradiation on the optical properties of bismuth doped hafnia-yttria-alumina-silicate fiber. Optical Materials Express. 2018;**8**:2550-2558

[7] Zhao Q, Luo Y, Hao Q, Peng GD. Electron beam irradiation and thermal-induced effects on the spectral properties of BAC-Al in Bi/Er codoped aluminosilicate fibers. Optical Materials Express. 2019;**9**:4287-4294

[8] Firstov SV, Khopin VF, Bufetov IA, Firstova EG, Guryanov AN, Dianov EM. Combined excitation-emission spectroscopy of bismuth active centers in optical fibers. Optics Express. 2011;**19**:19551-19561

[9] Obara K, Kakudate S, Oka K, Tada E, Morita Y, Sei M. Development of optical components for in-vessel viewing systems used for fusion experimental reactor. SPIE Proceedings. 1994;**2425**:115-122

[10] Paul SF, Goldstein JL, Durst RD, Fonck RJ. Effect of high-energy neutron flux on fiber optics in an active diagnostic on TFTR. The Review of Scientific Instruments. 1995;**66**:1252-1255

[11] Halder A, Kir'yanov AV, Sekiya EH, Saito K. Fabrication and characterization of bismuth-doped germano-silicate and phospho-silicate fibers for VIS/NIR applications. Optical Materials Express. 2019;**9**:1815-1825

[12] Kir'yanov AV, Halder A, Barmenkov YO, Sekiya EH, Saito K. Discussion on Raleigh scattering as a dominant loss factor in VIS/NIR in bismuth-doped silicate fibers [invited]. Optical Materials Express. 2019;**9**:2817-2827

[13] Halder A, Kir'yanov AV, Saito K. Broadband visible-to-NIR ASE from bismuth doped optical fibers: A window towards lasing. Japanese Journal of Applied Physics. 2019;**9**:120915

[14] Kir'yanov AV, Halder A, Sekiya E, Saito K, Barmenkov OA, Minkovich VP, et al. Impact of electron irradiation

upon optical properties of bismuth/yttria codoped phosphosilicate fiber. Optics & Laser Technology. 2020;**128**:106245

[15] Di Francesca D, Li Vecchi G, Girard S, Morana A, Reghioua I, Alessi A, et al. Qualification and calibration of single-mode phosphosilicate optical fiber for dosimetry at CERN. IEEE Journal of Lightwave Technology. 2019;**37**:4643-4649

[16] Di Francesca, D., Girard, S., Agnello, S., Alessi, A., Marchandella, C., Pailett, P., Querdane, Y., Kadi, Y., Brugger, M., Boukenter, A., "Combined temperature radiation effects and influence of drawing conditions on phosphorous-doped optical fibers," Physica Status Solidi A 216, art. 1800553 (2019)

[17] Girard S, Ouerdane Y, Marchandella C, Boukenter A, Quenard S, Authier N. Feasibility of radiation dosimetry with phosphorous-doped optical fibers in the ultraviolet and visible domain. Journal of Non-Crystalline Solids. 2011;**357**:1871-1874

[18] Girard S, Marchandella C, Morana A, Perisse J, Di Francesca D, Macé J-R, et al. Combined high dose and temperature radiation effects on multimode silica-based optical fibers. IEEE Transactions on Nuclear Science. 2013;**60**:4305-4313

Section 3

Bismuth Halide Perovskite Solar Cells

Chapter 6

Bismuth Halide Perovskites for Photovoltaic Applications

Khursheed Ahmad

Abstract

In the last decade, energy crisis has become the most important topic for researchers. Energy requirements have increased drastically. To overcome the issue of energy crisis in near future, numerous efforts and sources have been developed. Therefore, solar energy has been considered the most promising energy source compared to other energy sources. There were different kinds of photovoltaic devices developed, but perovskite solar cells have been considered the most efficient and promising solar cell. The perovskite solar cells were invented in 2009 and crossed an excellent power conversion efficiency of 25%. However, it has a few major drawbacks, such as the presence of highly toxic lead (Pb) and poor stability. Hence, numerous efforts were made toward the replacement of Pb and highly stable perovskite solar cells in the last few years. Bismuth halide perovskite solar cell is one type of the replacement introduced to overcome these issues. In this chapter, I have reviewed the role of bismuth halide perovskite structures and their optoelectronic properties toward the development of perovskite solar cells.

Keywords: methyl ammonium bismuth halide, perovskites, light absorbers, photovoltaics, perovskite solar cells

1. Introduction

In the last few decades, researchers/scientists have focused to find out the most efficient and promising technology to fulfill the energy requirements globally [1–7]. There are various renewable energy sources, but solar energy is a never-ending source. Photovoltaic or solar cell is a device that converts solar energy into electrical energy. In 2009, perovskite solar cells (PSCs) were developed, which exhibited the good open circuit voltage with decent power conversion efficiency (PCE) [6]. The term "perovskite" was created for calcium titanate ($CaTiO_3$) by Russian mineralogist L.A. Perovski [6]. However, another class of perovskite has also been discovered with molecular formula of ABX_3, where A = organic or inorganic cation such as $CH_3NH_3^+$ or Cs^+, B=Pb^{2+} or Sn^{2+}, etc. and X = halide anions. Miyasaka and co-workers have prepared methyl ammonium lead halide ($MAPbX_3$; MA = $CH_3NH_3^+$, X = I^- or Br^-) perovskites and used as visible light sensitizer in dye-sensitized solar cells (DSSCs) [6]. The fabricated device showed an interesting PCE of 3%. However, the use of liquid electrolyte destroyed its photovoltaic performance due to the dissolution of $MAPbX_3$ in the liquid electrolyte. Later, various strategies were developed to solve this issue, and a solid-state hole transport material was employed to overcome the dissolution of perovskites [8–19]. Later, some other strategies were made to further enhance the photovoltaic parameters of the perovskite solar cells. Lee et al. in 2012 reported an improved PCE of 10.9% by introducing solid-state perovskite solar cells [20]. In a short span of time, perovskite solar cells attracted researchers globally. In 2019, the highest PCE of 23.3% was reported for Pb halide perovskite solar cells [21]. This enhanced PCE is really impressive and has the potential for practical applications of perovskite solar cells. $MAPbX_3$ has also been used in other applications such as photodetectors, light-emitting-diodes, batteries and supercapacitors due to the presence of excellent optoelectronic properties in the perovskite materials [22–26]. $MAPbX_3$ has been considered the key material for the development of high-performance perovskite solar cells due to its excellent absorption coefficient. $MAPbX_3$ perovskite worked as a light absorber in perovskite solar cells. However, the presence of highly toxic Pb in the $MAPbX_3$ structure is a major drawback and a challenge for the scientific community to replace it with less toxic or nontoxic element. The presence of toxic Pb also restricted the practical use of perovskite solar cells. Thus, in the last 5 years, Pb has been replaced with Sn and Ge and showed good performance, but poor stability dismissed their performance [27, 28]. The perovskite structures possess general formula of ABX_3 as discussed above. The optical properties of the perovskite materials can be tuned by changing the A, B or X ion present in the ABX_3 structure. The size of the A, B or X should satisfy the Goldschmidt tolerance factor (t) equation given below:

$$(\mathrm{t}) = \frac{(rA + rX)}{\sqrt{2}(rB + rX.)} \tag{1}$$

where rA and rB are ionic radii of the A and B, while rX is the ionic radii of the X present in the ABX_3 structure. The perovskite materials showed good stability when tolerance factor is equal to 1. Recently, bismuth (Bi), antimony and copper, which are less toxic metals, have been used for the fabrication of Pb-free perovskite solar cells [29–32]. The Bi is a nontoxic element and has the similar properties and ionic size to those of Pb. The ionic radii of the Bi also satisfy the tolerance factor rule and enhanced the stability of the Bi-based perovskite materials. Moreover, it was found that Bi-based perovskite materials possess higher absorption coefficient, which makes them an efficient light-absorbing material for solar cell applications. Thus, Johansson et al. employed a ternary Bi-based perovskites with a molecular formula of $A_3Bi_2I_9$ (A = MA^+ or Cs^+) for Pb-free perovskite solar cells [33]. Buonassisi et al. also employed solvent engineering approach using $A_3Bi_2I_9$ perovskite material and the PCE of 0.71% was obtained [34]. Later, various approaches have been made to improve the performance of the Bi-based perovskite solar cells, and the highest PCE of 3.17% was achieved by applying vapor deposition approach [35]. In this chapter, I have reviewed the fabrication of perovskite solar cells. Recent advances in Bi-based perovskite solar cells and future prospective have also been described.

2. Fabrication of perovskite solar cells

Generally, perovskite solar cells are composed of different components such as electrode substrates (usually indium-doped tin oxide [ITI] or fluorine-doped tin oxide [FTO] glass substrates), perovskite light absorber layer, electron transport layer, hole transport layer and metal contact layer. Firstly, FTO or ITO glass substrates were cleaned and patterned with zinc powder and 2 molar hydrochloride solution to avoid short circuit in the device. Further, an electron transport layer was spin coated and annealed at ~500°C. Further, perovskite layer was deposited and annealed at temperature of ~ 70–120°C. Later, a hole transport material (HTM) layer was deposited followed by the deposition of metal contact layer of Au or Ag using thermal evaporation. The fabrication of perovskite solar cells has been illustrated in **Figure 1**.

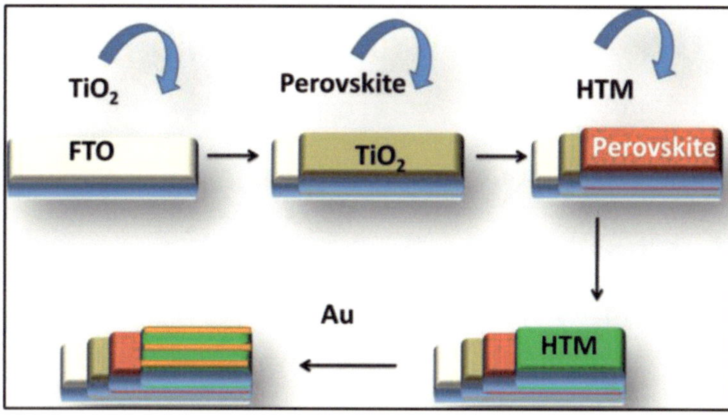

Figure 1.
Schematic illustration for the fabrication of perovskite solar cells.

The performance of the fabricated perovskite solar cells is investigated by various techniques such as photocurrent-voltage (I-V), external quantum efficiency (EQE), incident-photon-to-current-conversion efficiency (IPCE), photoluminescence spectroscopy, etc. Generally, the photovoltaic performance of any perovskite solar cells is estimated in terms of power conversion efficiency (PCE), fill factor, open circuit voltage and photocurrent density using I-V measurements. Photoluminescence spectroscopy revealed the electron lifetime of the generated electron in the perovskite structure, which is used to check the recombination reactions. It is believed that the perovskite solar cells with lower recombination reaction rate or high electron lifetime provide better performance in terms of efficiency.

3. Origin of bismuth halide perovskite solar cells

Since 2009, perovskite solar cells attracted enormous attention due to their excellent performance and simple fabrication process. However, the presence of highly toxic Pb remains a challenge. In recent years, bismuth has been employed to replace Pb from perovskite solar cells. Bismuth has been explored and Pb-free perovskite structures have been synthesized and employed for the development of Pb-free perovskite solar cells. Ahmad et al. employed new Pb-free perovskite light absorbers for perovskite solar cell applications [29–31]. The performance of these perovskite solar cells was less than 1%. In 2019, Lan et al. used $FA_3Bi_2I_9$ ($FA = CH(NH_2)_2$) perovskite material for perovskite solar cell applications [36]. In this work, the authors also prepared $MA_3Bi_2I_9$ ($MA = CH_3NH_3^+$) perovskite for perovskite solar cell applications. The X-ray diffraction (XRD) patterns of the $FA_3Bi_2I_9$ and $FA_3Bi_2I_9$ were found to be well-matched with the previous Joint Committee on Powder Diffraction Standards (JCPDS) data.

Ghasemi et al. [37] also prepared a new bismuth-based perovskite material phenethylammonium bismuth halides ($(PEA)_3Bi_2I_9$, $(PEA)_3Bi_2Br_9$ and $(PEA)_4Bi_2Cl_{10}$). The authors have prepared the crystal structures of these perovskite materials under facile conditions. The crystal structures of the $(PEA)_3Bi_2I_9$, $(PEA)_3Bi_2Br_9$ and $(PEA)_4Bi_2Cl_{10}$ have been presented in **Figure 2**. Furthermore, they have been investigated for their physiochemical properties for optoelectronic applications.

The crystal structures of $(PEA)_3Bi_2I_9$, $(PEA)_3Bi_2Br_9$ and $(PEA)_4Bi_2Cl_{10}$ showed the monoclinic crystal system with space group of $P2_1/n$, $P2_1/n$ and $P2_1/c$, respectively.

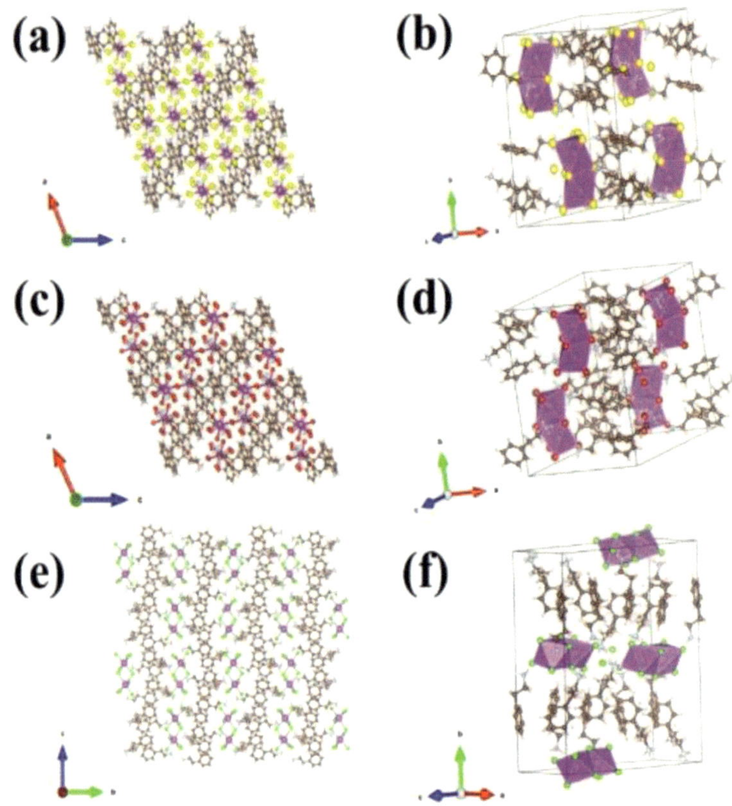

Figure 2.
Crystal structures of $(PEA)_3Bi_2I_9$ (a and b), $(PEA)_3Bi_2Br_9$ (c and d) and $(PEA)_4Bi_2Cl_{10}$ (e and f). Reproduced with permission [37].

These perovskite structures were composed of octahedral halide anionic clusters associated with PEA^+ cationic groups. In the crystal structures, Bi atom is situated in the center of each octahedron, whereas halide anions are in the corner of every face-sharing octahedron. The prepared perovskite structures possess zero-dimensional (0D) structure. The calculated XRD of $(PEA)_3Bi_2I_9$ from the single crystal structure and measured XRD of the powder sample have been presented in **Figure 3a**. The measured XRD of $(PEA)_3Bi_2I_9$ was well-matched with the calculated XRD pattern. The UV-vis spectra of the $(PEA)_3Bi_2I_9$, $(PEA)_3Bi_2Br_9$ and (PEA) have been plotted in **Figure 3b**. The band gap of the $(PEA)_3Bi_2I_9$, $(PEA)_3Bi_2Br_9$ and $(PEA)_4Bi_2Cl_{10}$ was calculated by Tauc relation as shown in **Figure 3c**. The direct band gap of the prepared $(PEA)_3Bi_2I_9$, $(PEA)_3Bi_2Br_9$ and $(PEA)_4Bi_2Cl_{10}$ perovskite structures was found to be 2.23, 2.66 and 3.28 eV, whereas indirect band gap was 2.38, 2.79 and 3.38 eV, respectively.

The UPS of the $(PEA)_3Bi_2I_9$ perovskite structure has been presented in **Figure 3d**, which provides better information of the energy levels of the $(PEA)_3Bi_2I_9$ perovskite structure. Further, perovskite solar cells were fabricated and cross-sectional scanning electron microscopic (SEM) image of the device has been presented in **Figure 4a**, whereas the photocurrent density (I)-voltage (V) curves have been depicted in **Figure 4b**. The SEM image showed the connection between the interlayers of the perovskite solar cell components. The I-V curves of the fabricated perovskite solar cells with $(PEA)_3Bi_2I_9$ and $(MA)_3Bi_2I_9$ perovskite light absorbers showed good open circuit voltage, but the performance was poor. The open circuit voltage of the $(PEA)_3Bi_2I_9$-based perovskite solar cell was high with improved

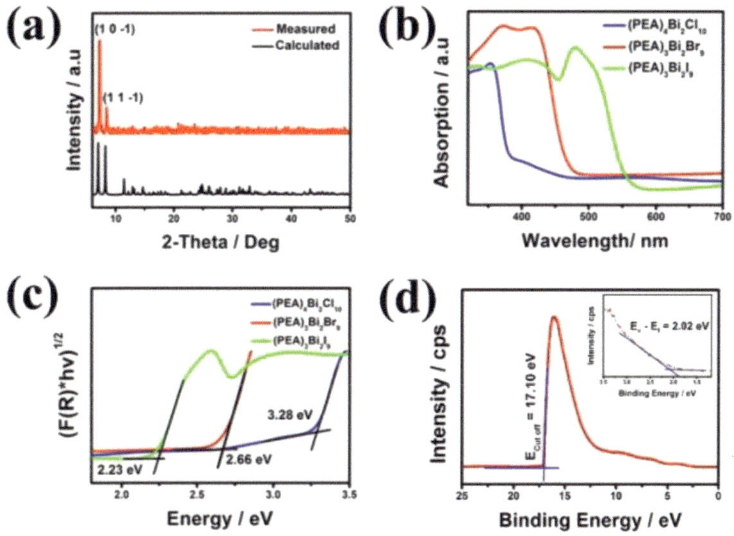

Figure 3.
Calculated and experimental XRD (a) of $(PEA)_3Bi_2I_9$, UV–vis spectra of $(PEA)_3Bi_2I_9$, $(PEA)_3Bi_2Br_9$ and $(PEA)_4Bi_2Cl_{10}$ (b), Tauc relation of $(PEA)_3Bi_2I_9$, $(PEA)_3Bi_2Br_9$ and $(PEA)_4Bi_2Cl_{10}$ (c) and ultraviolet photoemission spectroscopic (UPS) curve of $(PEA)_3Bi_2I_9$ (d). Reproduced with permission [37].

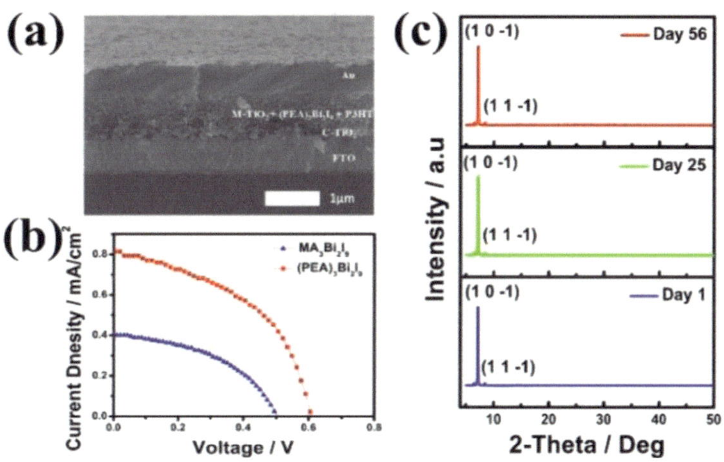

Figure 4.
Cross-sectional SEM images of perovskite solar cell device (a), I-V curve (b) of perovskite solar cells and XRD patterns of the $(PEA)_3Bi_2I_9$ perovskite structure exposed to air for different time. Reproduced with permission [37].

photocurrent density compared to the $(MA)_3Bi_2I_9$-based perovskite solar cells. The PCE of the $(PEA)_3Bi_2I_9$-based perovskite solar cells were found to be 0.23% with high open circuit voltage of 614 mV. However, the open circuit voltage for the $(MA)_3Bi_2I_9$-based perovskite solar cell was found to be 515 mV with PCE of 0.09%. The obtained results were found to be poor in terms of PCE.

Kulkarni et al. [38] also developed the Pb-free perovskite solar cells using $(MA)_3Bi_2I_9$ perovskite structure with novel strategies. Kulkarni et al. have employed N-methyl pyrrolidone (NMP) additive as morphology controller. In this work, different amount of NMP was added to observe the effect of NMP on the surface morphology of $(MA)_3Bi_2I_9$ perovskite.

The XRD patterns of the $(MA)_3Bi_2I_9$ without and with different amount of NMP were recorded and the obtained results have been presented in **Figure 5**. The obtained XRD patterned was well-matched with previously reported JCPDS data. This confirms the formation of $(MA)_3Bi_2I_9$ phase. However, the addition of NMP increases the peak intensity, which resulted in the orientation change. The other parameters also suggested that addition of NMP increases the crystallinity of the $(MA)_3Bi_2I_9$. Therefore, it can be said that the prepared $(MA)_3Bi_2I_9$ perovskite in the absence of NMP has poor crystalline phase or poor crystallinity. Furthermore, to investigate the effect of NMP on the morphological features, SEM images were recorded and have been presented in **Figure 6**.

The SEM image of $(MA)_3Bi_2I_9$ without NMP (**Figure 6a**) showed the hexagonal flakes with high exposure of TiO_2 and nonuniform surface. However, in case of $(MA)_3Bi_2I_9$ with NMP 12.5% (**Figure 6b**) showed large crystals, but poor uniform surface was observed. When 25% NMP (**Figure 6c**) was added, the surface morphology further changed. And the addition of 50% NMP drastically changed the surface morphology with larger grains as shown in **Figure 6d**. This suggested that 12.5% NMP was not sufficient to control the surface morphology of the $(MA)_3Bi_2I_9$. Moreover, $(MA)_3Bi_2I_9$ with 25% NMP covers the whole surface of the electrode substrate uniformly.

Further, perovskite solar cells were fabricated using $(MA)_3Bi_2I_9$ without and with NMP (at different percentage) under same conditions. The I-V and IPCE curves of the fabricated perovskite solar cells have been presented in **Figure 7a** and **b** respectively. The photocurrent density (J_{sc}) and PCE histograms were also presented in **Figure 7c** and **d**, respectively. The perovskite solar cell device fabricated in the absence of NMP showed PCE of 0.19% and open circuit voltage of 530 mV. However, the improved PCE of 0.31% was obtained for the perovskite solar cells fabricated with the addition of 25% NMP. In other cases, perovskite solar cells fabricated with 12.5% and 50% NMP showed lower performance, which may be due to the poor coverage of surface morphology.

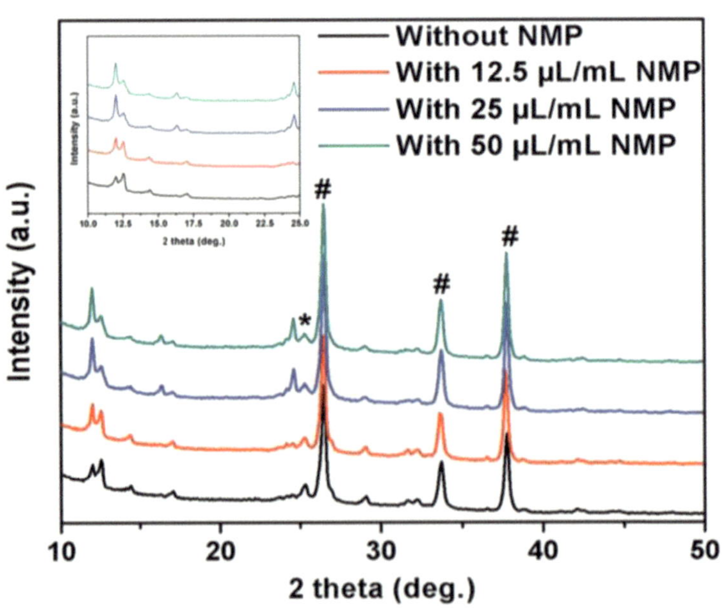

Figure 5.
*XRD of $(MA)_3Bi_2I_9$ without and with different amount of NMP. # peak for FTO and * for TiO_2. Reproduced with permission [38].*

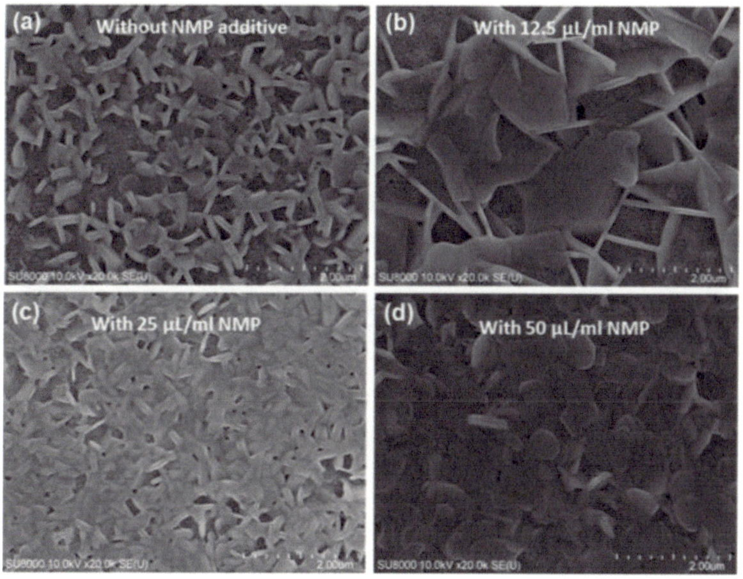

Figure 6.
SEM image of $(MA)_3Bi_2I_9$ without (a) and with different amount of NMP (b–d). Reproduced with permission [38].

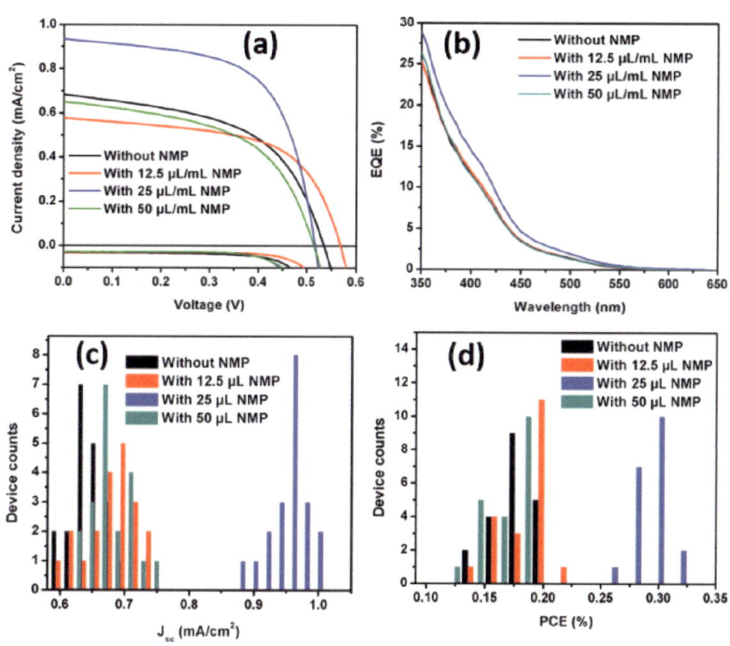

Figure 7.
Average I-V curves (a), IPCE (b), J_{sc} (c) and PCE (d) histogram of devices with and without NMP. Reproduced with permission [38].

In another work, Yu et al. [39] have prepared new perovskite solar cell device architecture of perovskite solar cells with mixed halide bismuth-based perovskite light absorber.

The authors have prepared a series of mixed halide-based perovskite structures. They have employed novel and simple strategies to prepare the Pb-free perovskite solar cells. The optical images of the $Cs_3Bi_2I_{9-x}Br_x$ perovskite series (x = 0, 1, 2, 3, 4, 6 and 9) have been presented in **Figure 8a**, which show the different colors of the prepared thin films, and this change in the color of the perovskite thin film was due to the change in the halide composition. Further, optical properties were checked by recording absorption spectra of the prepared thin films of $Cs_3Bi_2I_{9-x}Br_x$ perovskite series (x = 0, 1, 2, 3, 4, 6 and 9). The band gap of the $Cs_3Bi_2I_9$ was lower than that of the $Cs_3Bi_2Br_9$ as confirmed by the absorption spectra (**Figure 8b**).

The XRD patterns of the $Cs_3Bi_2I_{9-x}Br_x$ perovskite series (x = 0, 1, 2, 3, 4, 6 and 9) have been presented in **Figure 8c**. The obtained results showed strong and well-defined diffraction peaks, which suggested the high crystalline nature of the prepared Pb-free perovskite structures. The obtained XRD results were well-matched with previous JCPDS card number 23-0847. This confirms the successful formation of the prepared Pb-free perovskite structures. Further, perovskite solar cell devices were fabricated with these perovskite light absorbers. The schematic graph for the energy level values of the Pb-free perovskite solar cell components has been displayed in **Figure 9a**.

The schematic diagram showing that the electrons generated in the perovskite structure is transferred to the hole transport layers. The cross-sectional SEM image of the fabricated perovskite solar cells showed all the component layers of the perovskite solar cells in **Figure 9b**. Further, perovskite solar cells were fabricated by using different light absorbers $Cs_3Bi_2I_9$ and $Cs_3Bi_2I_6Br_3$. The I-V measurements (**Figure 10a**) and external quantum efficiency (EQE) were measured (**Figure 10b**) of the fabricated perovskite solar cells. The I-V curve of the $Cs_3Bi_2I_9$ showed poor performance, whereas the I-V curve of $Cs_3Bi_2I_6Br_3$ showed improved performance. The highest PCE of 1.15% was achieved for the $Cs_3Bi_2I_6Br_3$-based perovskite solar cells. The integrated J_{sc} for the $Cs_3Bi_2I_6Br_3$-based perovskite solar cells has been presented in **Figure 10b**, which showed the J_{sc} value of 3.11 mA/cm^2. However, the PCE of 0.23% was obtained for the $Cs_3Bi_2I_9$-based perovskite solar cell device. The enhanced performance of the $Cs_3Bi_2I_6Br_3$-based perovskite solar cells attributed to the lower band gap and higher absorption property of $Cs_3Bi_2I_6Br_3$ perovskite structure.

These obtained PCE were quite interesting, which is believed due to the tuning of the perovskite structure and better absorption activity of the $Cs_3Bi_2I_6Br_3$. Further improvements are still required to enhance the performance of the bismuth-based perovskite solar cells.

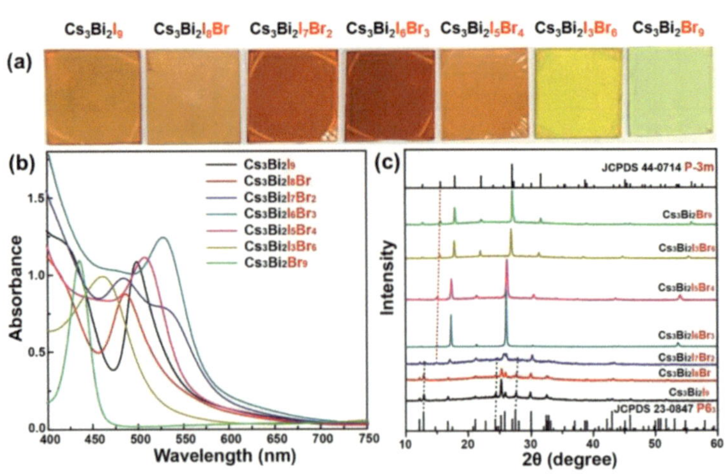

Figure 8.
Optical images (a), UV-vis spectra (b) and XRD patterns of $Cs_3Bi_2I_{9-x}Br_x$ (x = 0, 1, 2, 3, 4, 6 and 9). Reproduced with permission [39].

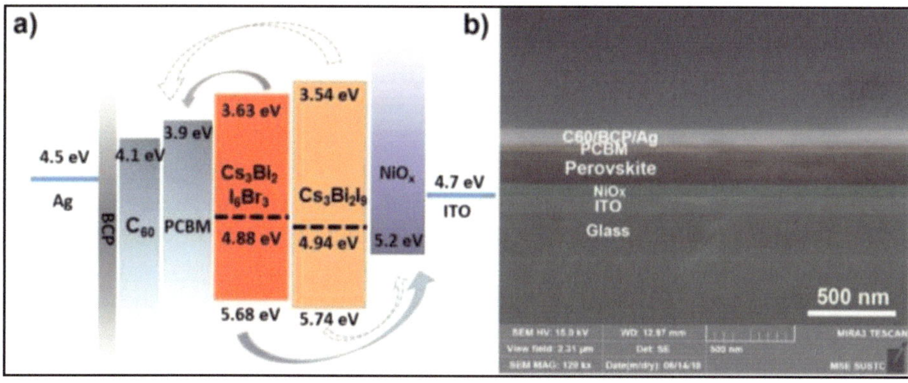

Figure 9.
Energy level graph (a) and cross-sectional SEM image (b) of perovskite solar cells. Reproduced with permission [39].

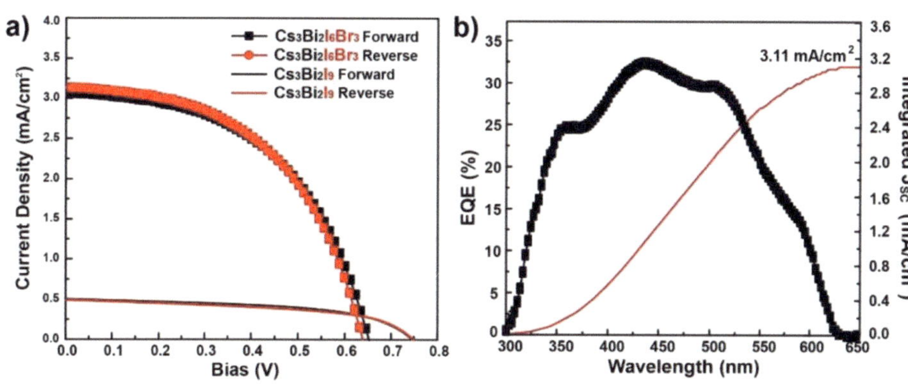

Figure 10.
I-V curves (a) and EQE (b) of the fabricated perovskite solar cells. Reproduced with permission [39].

4. Future prospective

It is well understood that Pb-based solar cells could not be commercialized due to the poor stability and toxic nature of Pb. Thus, Bi-based perovskite solar cells were developed under benign approaches as listed in **Table 1**. Since 2017, various approaches were made to construct the highly stable Pb-free PSCs. In this regard, Ghasemiet et al. [37] prepared $(PEA)_3Bi_2I_9$ perovskite light absorber and constructed the PSC device. The developed device exhibited PCE of 0.3%. Kulkarni

S. No.	Light absorbers	J_{sc} (mA/cm^2)	V_{oc} (mV)	PCE (%)	References
1.	$(PEA)_3Bi_2I_9$	~0.81	614	0.3	[37]
2.	$(MA)_3Bi_2I_9$	~0.4	515	0.09	[37]
3.	$(MA)_3Bi_2I_9$	~0.87	530	0.3	[38]
4.	$Cs_3Bi_2I_6Br_3$	3.11	~650	1.15	[39]
5.	$Cs_3Bi_2I_9$	~0.501	~750	0.23	[39]

Table 1.
Photovoltaic parameters of Bi-based perovskite solar cells.

et al. [38] also employed novel and unique approaches to improve the surface morphology of the $(MA)_3Bi_2I_9$ perovskite film, and the developed PSC device showed PCE of 0.3%. Yu et al. [39] also tuned the optical properties of the $Cs_3Bi_2I_9$ perovskite structure. Thus, the designed and synthesized $Cs_3Bi_2I_6Br_3$ perovskite material-based PSC device showed enhanced PCE of 1.15%. These obtained results showed the excellent optical properties of the Bi-based perovskite materials and suggested their potential as light absorbers in the construction of PSCs.

The Bi-based perovskite structures have wide band gap ~2.2 eV, which limits their absorption properties. Moreover, poor surface morphology of the Bi-based perovskite structures has been the other drawback, which resulted in the poor performance. Hence, the following strategies could be the key points to further improve the performance of the Bi-based perovskite solar cells:

1. Introducing new device architecture of the perovskite solar cells may improve the performance of the perovskite solar cells.

2. The performance of the Bi-based perovskite solar cells could be enhanced by developing the novel charge extraction/electron transport layer.

3. Some new approaches can be applied to prepare the high-quality thin films of the Bi-based perovskite structures, which further help to improve the photovoltaic parameters.

4. Solvent engineering and doping of Bi-based perovskite structure may also improve the performance of the perovskite solar cells.

5. Conclusions

The origin and advances in the field of perovskite solar cells have been reviewed. Perovskite solar cells could be the future energy source with low cost. The replacements of the toxic Pb with different metals have been investigated to produce Pb-free perovskite structures for the development of Pb-free perovskite solar cells. Bismuth, which is a nontoxic element, has been widely employed for the development of highly stable and Pb-free perovskite structures. Bi-based perovskite structures have shown promising performance and their performance may be further improved by incorporating unique approaches and efficient charge extraction layers.

Acknowledgements

K.A. sincerely acknowledged Discipline of Chemistry, IIT Indore. K.A. also thanks UGC, New Delhi, India, for RGNFD fellowship.

Conflict of interest

The author declares no conflict of interest.

Author details

Khursheed Ahmad
Discipline of Chemistry, Indian Institute of Technology Indore, Simrol, M.P., India

*Address all correspondence to: khursheed.energy@gmail.com

References

[1] Sum TC, Mathews N. Advancements in perovskite solar cells: Photophysics behind the photovoltaics. Energy & Environmental Science. 2014;**7**:2518-2534

[2] Reddy VS, Kaushik SC, Ranjan KR, Tyagi SK. State-of-the-art of solar thermal power plants. Renewable and Sustainable Energy Reviews. 2013;**27**:258-273

[3] Kitano M, Hara M. Heterogeneous photocatalytic cleavage of water. Journal of Materials Chemistry. 2010;**20**:627-641

[4] Ahmad K, Mobin SM. Graphene oxide based planar heterojunction perovskite solar cell under ambient condition. New Journal of Chemistry. 2017;**41**:14253-14258

[5] Chen Y-Z, Wu R-J, Lin LY, Chang WC. Novel synthesis of popcorn-like TiO2 light scatterers using a facile solution method for efficient dye-sensitized solar cells. Journal of Power Sources. 2019;**413**:384-390

[6] Kojima A, Teshima K, Shirai Y, Miyasaka T. Organometal halide perovskites as visible-light sensitizers for photovoltaic cells. Journal of the American Chemical Society. 2009;**131**:6050-6051

[7] Ahmad K, Mohammad A, Mobin SM. Hydrothermally grown α-MnO2 nanorods as highly efficient low cost counter-electrode material for dye-sensitized solar cells and electrochemical sensing applications. Electrochimica Acta. 2017;**252**:549-557

[8] Im JH, Lee CR, Lee JW, Park SW, Park NG. 6.5% efficient perovskite quantum-dot-sensitized solar cell. Nanoscale. 2011;**3**:4088-4093

[9] Kim HS, Lee CR, Im JH, Lee KB, Moehl T, Marchioro A, et al. Lead iodide perovskite sensitized all-solid-state submicron thin film mesoscopic solar cell with efficiency exceeding 9%. Scientific Reports. 2012;**2**:591

[10] Wehrenfennig C, Liu M, Snaith HJ, Johnston MB, Herz LM. Charge-carrier dynamics in vapour-deposited films of the organolead halide perovskite CH3NH3PbI3−xClx. Energy & Environmental Science. 2014;**7**:2269-2275

[11] Ma J, Guo X, Zhou L, Lin Z, Zhang C, Yang Z, et al. Enhanced planar perovskite solar cell performance via contact passivation of TiO2/perovskite interface with NaCl doping approach. ACS Applied Energy Materials. 2018;**1**:3826-3834

[12] Ke W, Fang G, Wang J, Qin P, Tao H, Lei H, et al. Perovskite solar cell with an efficient TiO2 compact film. ACS Applied Materials & Interfaces. 2014;**6**:15959-15965

[13] Guo Z, Liguo G, Zhang C, Xu Z, Ma T. Low-temperature processed non-TiO2 electron selective layers for perovskite solar cells. Journal of Materials Chemistry A. 2018;**6**:4572-4589

[14] Peng G, Wu J, Wu S, Xu X, Ellis JE, Xu G, et al. Perovskite solar cells based on bottom-fused TiO2 nanocones. Journal of Materials Chemistry A. 2016;**4**:1520-1530

[15] Lv M, Lv W, Fang X, Sun P, Lin B, Zhang S, et al. Performance enhancement of perovskite solar cells with a modified TiO2 electron transport layer using Zn-based additives. RSC Advances. 2016;**6**:35044-35050

[16] Liu D, Kelly TL. Perovskite solar cells with a planar heterojunction structure prepared using room-temperature solution processing techniques. Nature Photonics. 2014;**8**:133-138

[17] Jeong S, Seo S, Park H, Shin H. Atomic layer deposition of a SnO2 electron-transporting layer for planar perovskite solar cells with a power conversion efficiency of 18.3%. Chemical Communications. 2019;**55**:2433-2436

[18] Wang S, Zhu Y, Liu B, Wang C, Ma R. Introduction of carbon nanodots into SnO2 electron transport layer for efficient and UV stable planar perovskite solar cells. Journal of Materials Chemistry. 2019;7:5353-5362

[19] Ding B, Huang SY, Chu QQ, Li Y, Li CX, Li CJ, et al. Low-temperature SnO2-modified TiO2 yields record efficiency for normal planar perovskite solar modules. Journal of Materials Chemistry A. 2018;**6**:10233-10242

[20] Lee MM, Teuscher J, Miyasaka T, Murakami TN, Snaith HJ. Efficient hybrid solar cells based on meso-superstructured organometal halide perovskites. Science. 2012;**338**:643-647

[21] Zimmermann I, Aghazad S, Nazeeruddin MK. Lead and HTM free stable two-dimensional tin perovskites with suitable band gap for solar cell applications. Angewandte Chemie, International Edition. 2019;**131**:1084-1088

[22] He M, Chen Y, Liu H, Wang J, Fang X, Liang Z. Chemical decoration of $CH_3NH_3PbI_3$ perovskites with graphene oxides for photodetector applications. Chemical Communications. 2015;**51**:9659-9661

[23] Xu JT, Chen YH, Dai LM. Efficiently photo-charging lithium-ion battery by perovskite solar cell. Nature Communications. 2015;**6**:8103

[24] Tathavadekar M, Krishnamurthy S, Banerjee A, Nagane S, Gawli Y, Suryawanshi A, et al. Low-dimensional hybrid perovskites as high performance anodes for alkali-ion batteries. Journal of Materials Chemistry A. 2017;**5**:18634-18642

[25] Leyden MR, Meng L, Jiang Y, Ono L, Qiu EJ, Juarez-Perez C, et al. Methylammonium lead bromide perovskite light-emitting diodes by chemical vapor deposition. Journal of Physical Chemistry Letters. 2017;**8**:3193-3198

[26] Xu X, Li S, Zhang H, Shen Y, Zakeeruddin SM, Graetzel M, et al. A power pack based on organometallic perovskite solar cell and supercapacitor. ACS Nano. 2015;**9**:1782-1787

[27] Krishnamoorthy T, Ding H, Yan C, Leong WL, Baikie T, Zhang Z, et al. Lead-free germanium iodide perovskite materials for photovoltaic application. Journal of Materials Chemistry A. 2015;**3**:23829-23832

[28] Koh TM, Krishnamoorthy T, Yantara N, Shi C, Leong WL, Boix PP, et al. Formamidinium tin-based perovskite with low Eg for photovoltaic applications. Journal of Materials Chemistry A. 2015;**3**:14996-15000

[29] Ahmad K, Kumar P, Mobin SM. A two-step modified sequential deposition method-based Pb-free (CH3NH3)3Sb2I9 perovskite with improved open circuit voltage and performance. ChemElectroChem. 2020;7:946-950

[30] Ahmad K, Ansari SN, Natarajan K, Mobin SM. A two-step modified deposition method based (CH3NH3)3Bi2I9 perovskite: Lead free, highly stable and enhanced photovoltaic performance. ChemElectroChem. 2019;**6**:1-8

[31] Ahmad K, Ansari SN, Natarajan K, Mobin SM. Design and Synthesis of 1D-Polymeric Chain Based [(CH3NH3)3Bi2Cl9]n Perovskite: A New Light Absorber Material for Lead Free Perovskite Solar Cells. ACS Applied Energy Materials. 2018;**01**:2405-2409

[32] Ahmad K, Mobin SM. Organic–Inorganic Copper (II)-Based

Perovskites: A Benign Approach toward Low-Toxicity and Water-Stable Light Absorbers for Photovoltaic Applications. Energy Technology. 2019. DOI: 10.1002/ente.201901185

[33] Park BW, Philippe B, Zhang X, Rensmo H, Boschloo G, Johansson EMJ. Bismuth based hybrid perovskites $A_3Bi_2I_9$ (a: Methylammonium or cesium) for solar cell application. Advanced Materials. 2015;**27**:6806

[34] Shin SS, Correa Baena JP, Kurchin RC, Polizzotti A, Yoo JJ, Wieghold S, et al. Solvent-engineering method to deposit compact bismuth-based thin films: Mechanism and application to photovoltaics. Chemistry of Materials. 2018;**30**:336-343

[35] Jain SM, Phuyal D, Davies ML, Li M, Philippe B, De Castro C, et al. An effective approach of vapour assisted morphological tailoring for reducing metal defect sites in lead-free, (CH3NH3)(3)Bi2I9 bismuth-based perovskite solar cells for improved performance and long-term stability. Nano Energy. 2018;**49**:614-624

[36] Lan C, Liang G, Zhao S, Lan H, Peng H, Zhang D, et al. Lead-free formamidinium bismuth perovskites (FA)3Bi2I9 with low bandgap for potential photovoltaic application. Solar Energy. 2019;**177**:501-507

[37] Ghasemi M, Lyu M, Roknuzzaman M, Yun JH, Hao M, He D, et al. Phenethylammonium bismuth halides: from single crystals to bulky-organic cation promoted thin-film deposition for potential optoelectronic applications. Journal of Materials Chemistry A. 2019;**7**:20733-20741

[38] Kulkarni A, Singh T, Ikegami M, Miyasaka T. Photovoltaic enhancement of bismuth halide hybrid perovskite by N-methyl pyrrolidone-assisted morphology conversion. RSC Advances. 2017;**7**:9456-9460

[39] Yu BB, Li M, Yang J, Chen W, Zhu Y, Zhang X, et al. Alloy-induced phase transition and enhanced photovoltaic performance: The case of Cs3Bi2I9-xBrx perovskite solar cells. Journal of Materials Chemistry A. 2019;**7**:8818-8825

Section 4

Bismuth Related Data Storage Materials

Chapter 7

Synthesis and Characterization of Multiferroic $BiFeO_3$ for Data Storage

Kuldeep Chand Verma

Abstract

Multiferroic $BiFeO_3$ deals with spintronic devices involved spin-charge processes and applicable in new non-volatile memory devices to store information for computing performance and the magnetic random access memories storage. Since multiferroic leads to the new generation memory devices for which the data can be written electrically and read magnetically. The main advantage of present study of multiferroic $BiFeO_3$ is that to observe magnetoelectric effects at room temperature. The nanostructural growth (for both size and shape) of $BiFeO_3$ may depend on the selection of appropriate synthesis route, reaction conditions and heating processes. In pure $BiFeO_3$, the ferroelectricity is induced by $6s^2$ lone-pair electrons of Bi^{3+} ions and the G-type antiferromagnetic ordering resulting from Fe^{3+} spins order of cycloidal (62-64 nm wavelength) occurred below Neel temperature, T_N = 640 K. The multiferroicity of $BiFeO_3$ is disappeared due to factors such as impurity phases, leakage current and low value of magnetization. Therefore, to overcome such factors to get multiferroic enhancement in $BiFeO_3$, there are different possible ways like changes dopant ions and their concentrations, $BiFeO_3$ composites as well as thin films especially multilayers.

Keywords: electric-driven magnetic switching, chemical synthesis, magnetoelectric, ferroelectric polarization

1. Introduction

Spintronic devices that electrically store non-volatile information are the potential candidates for high-performance, high-density memories due to their interdependence between magnetization and charge transport phenomenon. Since the capacitor stored information in the form of charges and the electric field moved these charges to transmit information. However, the magnetic recording is caused when the magnetic field used to read or write the information stored in the form of magnetization by measuring local orientation of spins. The behavior started to change in 1988, when the discovery of giant magnetoresistance provides a way for efficient control of charge transport through magnetization [1, 2], for example, the hard-disk recording. Hard disk drives (HDD) with a capacity of 10 MB were sold for ~\$5300 in the 1980s, and were unaffordable for many during Apple and IBM PC era. However, HDDs with 16 TB capacity are available at the time of writing (2020). The computers in 1980s had memory of hundreds of kB that recently 8 GB random access memory (RAM). Even mobile gadgets have a dynamic random access memory (DRAM) capacity of ~4 GB, at the time of writing. This is possible by use of charge property of the electron as well the spintronics devices make use of the spin property of an electron. Flash memory is an example, as it is a non-volatile memory as used in mobile applications [3]. However, the electrically induced bistable magnetization switching at room temperature - a necessary requirement for magnetic data storage - is the multiferroics [4]. The room-temperature manipulation of magnetization by an electric field using the multiferroic $BiFeO_3$ represents an essential step toward magnetoelectric (ME) control for spintronics devices [5].

1.1 Multiferroic memories for data storage

Multiferroic might hold the future for the ultimate memory device. The demonstration of a four-state resistive memory element in a tunnel junction with multiferroic barriers represents a major step in this direction [6–9]. For example, the thin films of lanthanum bismuth manganite remain ferromagnetic and ferroelectric down to thicknesses of 2 nm and, when used as a multiferroic tunneling junction, act as a four-state resistive system. Their spin-filter device as shown in **Figure 1(a)** is the tunnel junctions, which has tunnel barrier height is spin dependent because the bottom level of the conduction band in the ferromagnetic barrier is spin-split by exchange model. This allows the tunneling of electrons that to be efficiently filtered according to their spin. Gajek M *et al.* [10] suggested the large tunnel magnetoresistance in junctions that have a ferromagnetic electrode. The combination of these two effects - magnetoresistance plus electroresistance - yields a four-state resistive memory element. In comparison to the information stored in a capacitor, the resistive memories, on the other hand, can be read more simply, for example, by monitoring the source-drain current in a field-effect transistor. In order to make a multiple-state ME memory, one must be able to access the four states formed by electric polarization P and magnetization M, *i.e.* (+P, +M), (+P, −M), (−P, +M), and (−P, −M) [6].

1.2 Spin-transfer-torque magnetic random access memory

The spin-transfer-torque magnetic random access memory (STT-MRAM) devices stored information due to use of magnetic orientation in the ferromagnetic

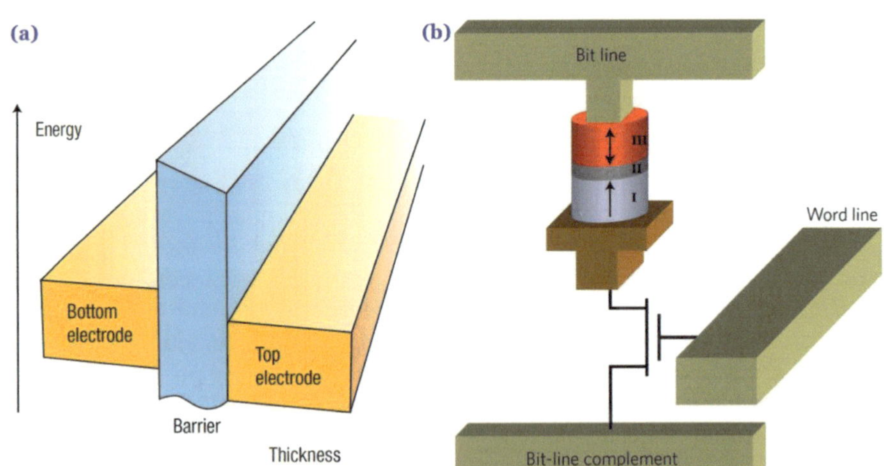

Figure 1.
(a) Tunnel junction (electrons tunnel from bottom electrode through barrier into top electrode): Schematic [6]. (b) Spin-transfer-torque magnetic random access memory (STT-MRAM) bit cell. A magnetic tunnel junction (I), a tunnel barrier (II) and a free-layer element (III), with both layers magnetized perpendicular to the plane of the junction (arrows). Bit is selected by a word line and transistor [7].

nanoparticles. For example, hard disk drives (use magnetic states to store information). In addition to hard disk drives, the STT-MRAM is a device that written and read electrically without any moving parts. The function of spin-transfer is to write information and such information is read by measuring the device resistance. The magnetoresistance plays role to measure percentage change in resistance between parallel and antiparallel magnetic spins of the electrodes of the magnetic tunnel junction. Such magnetic tunnel junction is made up by a ferromagnetic metal/insulator/ferromagnetic material [7]. **Figure 1(b)** shows a 1-bit STT-MRAM cell constituted by free layer and reference layer that are magnetized perpendicular to the plane of the junction [7]. The cell is constituted with a word line with a transistor that required for each cell. The biasing voltage could operate the cell with respect to bit lines and such read bias voltage measured the cell resistance to determine the bit state to be low of 100 mV. However, the write bias voltage is higher to allow the magnetic moment of the free layer taken to be reversed by using spin transfer torque.

1.3 Multiferroic $BiFeO_3$

The multiferroic $BiFeO_3$ (BFO) has high Curie temperature, $T_c \sim 1103$ K and Neel temperature, $T_N \sim 643$ K results into simultaneously ferroelectric and antiferromagnetic orders at room temperature. The ferroelectricity in BFO is originated by $6s^2$ lone pair electrons of Bi^{3+} via structural distortion, however, magnetism resulted with Fe-O-Fe superexchange interactions [11]. But the reported study pointed out BFO with low spontaneous polarization and saturation magnetization because superimposition of a spiral spins structure of BFO by antiferromagnetic order. In such spiral spin structure, the antiferromagnetic axis rotates BFO crystal with 62 nm long wavelength, which cancels out the macroscopic magnetization as well as affects ME coupling value. The superexchange between the octahedrally coordinated Fe^{3+} through the O ligand is responsible for the resulting antiferromagnetism. The presence of oxygen vacancies and the valence fluctuation (Fe^{2+}/Fe^{3+}) believed to be the main disadvantages causing large electrical leakage in BFO.

1.3.1 $BiFeO_3$ structure

The BFO structure is characterized with two distorted perovskite units connected by their body diagonal to build a rhombohedral unit cell as shown in **Figure 2(a)**. For such BFO structure, the two octahedral oxygen connected along (111) plane are rotated clockwise and counterclockwise at 13.8°. However, Fe ion is shifted 0.135 Å along same axis from oxygen that present at the octahedral position. The large displacement of Bi ions with respect to the FeO_6 responsible to induced ferroelectric polarization [12–14]. The spins in this BFO structure are incommensurate to form antiferromagnetic order. There is also some canting moments to give weak ferromagnetism due to Dzyaloshinskii-Moriya (DM) effect because the moments may oriented perpendicular to the (111) polarization direction to influence symmetry properties.

1.3.2 Lone-pair mechanism supporting multiferroicity

The spatial asymmetry that caused by anisotropic unbounded valence electrons around Bi^{3+} might to give lone pair mechanism (**Figure 2(b)**) responsible into room temperature ferroelectric polarization of BFO [12]. In $BiFeO_3$, a pair of Bi^{3+} valence electrons of the 6 *s* orbital not involved *sp* hybridization to generates a local dipole which resulting into $\sim$100 μC cm^{-2} ferroelectric polarization below $T_C = 1103$ K. A long-range periodic antiferromagnetic structure arises below $T_N = 643$ K. The main

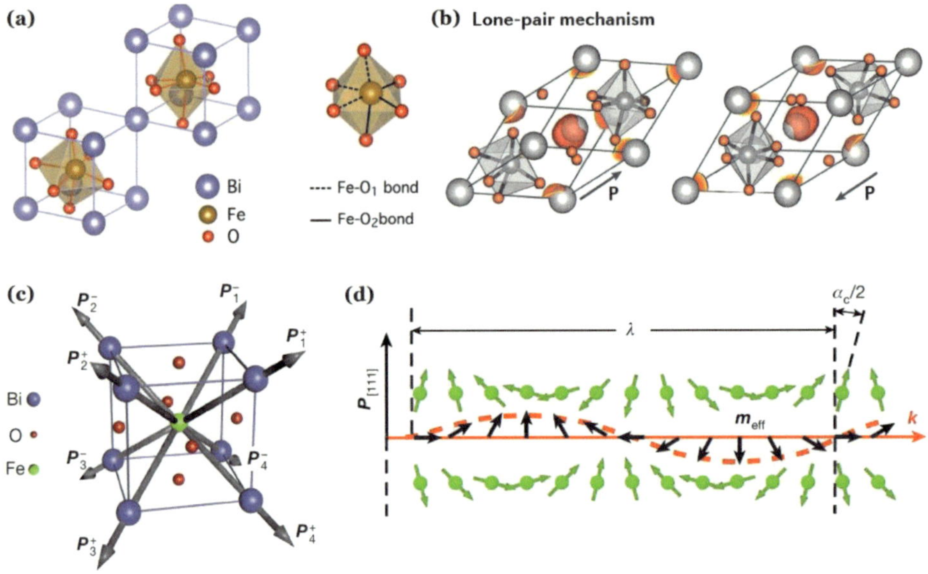

Figure 2.
(a) Rhombohedral (R3c) structure of $BiFeO_3$. (b) Ferroelectricity of $BiFeO_3$ due to lone-pair electrons (iso-surface: red) [12]. (c) Pseudocubic unit cell of $BiFeO_3$, (d) spin cycloid of canted antiferromagnetic alignments [13].

driving force of the ferroelectric phase transitions seems to be the stereochemical activity of the Bi 6 s lone-pair, resulting into a displacement of Bi and O sublattices. However, the Fe sublattice is also displaced and makes a sizeable contribution to the total electric polarization.

1.3.3 Ferroelectric and magnetic order in $BiFeO_3$

Bulk $BiFeO_3$ crystallized into a slightly distorted rhombohedral structure which commonly described by the pseudocubic unit cell (**Figure 2(c)**). The displacement of Bi ions relative to the FeO_6 octahedra gives rise to a strong ferroelectric polarization ($100\ \mu C\ cm^{-2}$) along one of the [111] directions [13]. However, the magnetization in $BiFeO_3$ involved G-type antiferromagnet order with a cycloidal wavelength, $\lambda \sim$ 62-64 nm was investigated by high-resolution neutron diffraction. As shown in **Figure 2(d)**, the normalized position of the spin cycloid propagation and the ferroelectric polarization vector might to induce ME coupling effect.

2. Synthesis of multiferroic $BiFeO_3$

The some synthesis methods used for $BiFeO_3$ are summarized in **Table 1**. Some of the listed methods are the derivatives of wet-chemical that deals with chemical reactions in the solution phase using precursors at proper stoichiometric conditions. Each wet-chemical synthesis route differs from the others in the sense that one cannot find a general rule for these kinds of synthesis approaches. Such wet synthesis schemes mostly used for fabrication of 2D nanomaterials. The wet-chemical processes offer a high degree of controllability and reproducibility of the 2D nanostructures. The solvothermal synthesis, template synthesis, self-assembly, oriented attachment, hot-injection, and interface-mediated synthesis are wet-chemical routes. However, the solvothermal and hydrothermal processes are mostly used to synthesize 2D nanostructures due to their simple and scalable steps.

Synthesis method	Reaction time	Precursor salts	Reaction condition	Shape control	Shape/size	Ref.
Wet chemical methods	Hours	$Bi(NO_3)_3.5H_2O$, $Fe(NO_3)_3$	HNO_3 used to adjust pH, Ethylene glycol & carboxylic acid are a polymerizing agent, annealed at 400°C/2 h	Good	13-70 nm particles	[15, 16]
Sol-gel method	Hours/ day	$Bi_5H_9N_4O_{22}$, $Fe(NO_3)_3$	Precursor concentration adjusted in 0.05-0.2 M using acetic acid & ethylene alcohol, annealed at 200-500°C	Good	10 nm particles	[17]
Co-precipitation method	Minutes	$Fe(NO_3)_3{\cdot}9H_2O$, Bi_2O_3	NaOH used as a precipitating agent, maintain pH 12, annealed at 400-600°C/1 h	Poor	200-250 nm particles	[18]
Hydrothermal process	Hours/ days	$Bi(NO_3)_3{\cdot}5H_2O$, $FeCl_3{\cdot}6H_2O$	Precursor salts dissolved in acetone, pH adjusted 10-11 by ammonia solution, Precursor solution transferred into teflon-lined steel autoclave and heated at 180°C for 72 h	Very good	45-200 nm diameter wires	[19]
Solution evaporation method	Hours	$Fe(NO_3)_3.9H_2O$, $Bi(NO_3)_3.5H_2O$, HNO_3	Tartaric acid and nitric acid used as precipitating and oxidizing agent, crystallization of the final powder take-place at 650°C/2 h	Good	22-31 nm particles	[20]
Microwave-assisted hydrothermal synthesis	Minutes	$Bi(NO_3)_3.5H_2O$, $Fe(NO_3)_3.9H_2O$	NaOH solution and polyethylene glycol were added to the precursors to obtain brown precipitates, solution irradiated by 300 W of MW irradiation for 30 min at 190°C with 2450 Hz	Good	20 nm diameter wires	[21, 22]
Self-catalyzed fast reaction process	Hours	α-Fe_2O_3, Bi_2O_3,	5 mol of tartaric acid ($C_4H_6O_6$) added to the precursors and heated at 250°C to begin to ignite and violently burn, final powder annealed at 650°C/2 h	Poor	100 nm particles	[23]
Conventional solid state	Hours/ days	Bi_2O_3, Fe_2O_3, Co_3O_4	Synthesis of $BiFeO_3$ powder from Bi_2O_3	Poor	100-200 nm particles	[24]

Synthesis method	Reaction time	Precursor salts	Reaction condition	Shape control	Shape/size	Ref.
reaction and thin film deposition of $BiFeO_3$-$CoFe_2O_4$			and Fe_2O_3 and heated at 800°C, $CoFe_2O_4$ powder prepared from Co_3O_4 and Fe_2O_3 using ball milling for 24 hours and heated at 1200°C/3 h, RF magnetron sputtering used for thin film deposition			
Chemical combustion method	Hours	$Bi(NO_3)_3$, $Fe(NO_3)_3$	Precursor solution mixed in polyethylene glycol, Urea added at 70°C and the combustion take-place, annealed at 600°C/5 h	Poor	50-500 nm nanostructures	[12]
Polymer-directed solvothermal	Hours	$Bi(NO_3)_3$ $.5H_2O$, $Fe(NO_3)_3$ $.9H_2O$	Precursor salt dissolved in HNO_3 and dipped by ethanol containing 1 g of PVP, and added 1.2 g of NaOH, transferred precursor solution into Teflon liner steel autoclave and heated at 180°C/6 h	Very good	1-D nanoparticles-assembled microrods	[25]
Sol-gel template process	Hours/ days	Bi $(NO_3)_3.5H_2O$, $Fe(NO_3)_3$. $9H_2O$	Porous nanochannel alumina (NCA) templates used, precursor salts mixed in nitric acid to get transparent, Citric acid & deionized water added, pH adjusted to be natural by using ammonia, urea added in 1/20th ratio, NCA templates added and heated at 80°C/20 h and annealed at 650° C/5 h	Very good	Nanotubes 150- 190 nm	[26]
Sonochemical technique	Hours	$Bi(NO_3)_3$ $.5H_2O$, Fe $(NO_3)_3$ $.9H_2O$, Mn $(OOCCH_3)_2$ $.4H_2O$, Cr $(NO_3)_3$ $.9H_2O$	In sonicated solution of Bi and Fe, add 5 ml of tetraethylene glycol and sonicated for 10 min, pH adjusted to 8 by adding ammonia and irradiated with a high intensity (100 W cm^{-2})	Good	Nanorods diameter 20-50 nm	[27]

Synthesis method	Reaction time	Precursor salts	Reaction condition	Shape control	Shape/size	Ref.
			ultrasonic radiation of 20 kHz, final product heated at 400°C/1 h			
Anodized alumina template technique	Hours	$Bi(NO_3)_3 \cdot 5H_2O$, $Fe(NO_3)_3 \cdot 9H_2O$	Precursor salts are mixed in 2-methoxyethanol, *p*H 4-5 by adding 2-methoxyethanol and nitric acid, anodized aluminum oxide template immersed in precursors for 12 h and heated at 750°C/12 h	Very good	Wires ~50 nm in diameter	[28]
Sol-gel based electrospinning	Hours	$Bi(NO_3)_3 .5H_2O$, $Fe(NO_3)_3 \cdot 9H_2O$	Precursors salts are neutralized with 2-methoxyethanol, pH 3-4 adjusted with Ethanolamine, Ethanol, glacial acetic acid, and poly vinyl pyrrolidone (PVP) added, solution was electrospun and the ultrafine fibers spun were collected in glass flake or Pt/Ti/SiO_2/Si substrate, final heating at 550°C/2 h	Very good	Nanofiber diameter in 100-300 nm	[29]

Table 1.
List the synthesis methods used to fabricate $BiFeO_3$.

3. Results and discussion

3.1 Structural analysis of $BiFeO_3$

3.1.1 X-ray diffraction of Pb substituted $BiFeO_3$

Figure 3(a) shows the X-ray diffraction (XRD) pattern of $Bi_{1-x}Pb_xFeO_3$ [x = 0 (BFO), 0.05 (BPFO5), 0.075 (BPFO75) and 0.1 (BPFO10)] nanostructures measured at room temperature [11]. All reflections are indexed to a rhombohedral structure of $R3c$ space group. This is based on the character of the single (012) peak at around 22° and the splitting of the (104) and (110) peaks around 32°. The splitting of XRD peaks indicate the structural distortion due to tilting of FeO_6 octahedrons. The calculated lattice constants are a(Å) = 5.578, 5.577, 5.573 and 5.574 and c(Å) = 13.862, 13.893, 13.905 and 13.915, respectively for BFO, BPFO5, BPFO75 and BPFO10. The increase in lattice constant, c and decrease in a may be due to change in Fe-O bond lengths and Fe-O-Fe bond angles which have a significant effect on multiferroic properties.

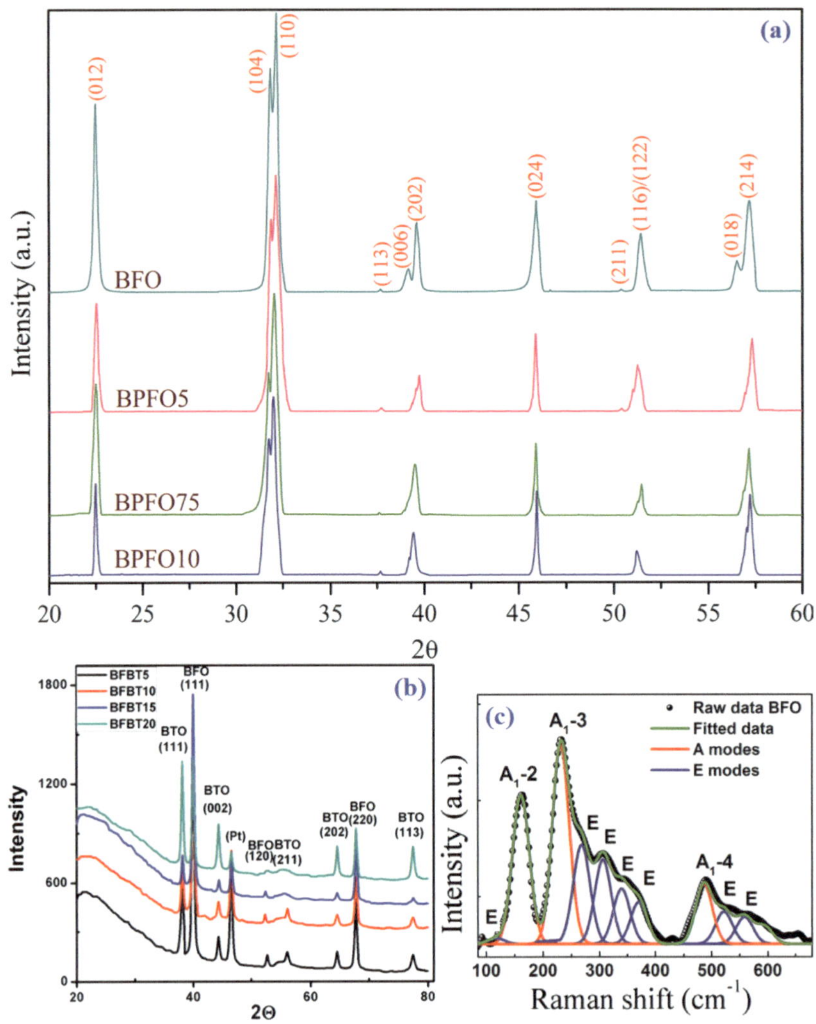

Figure 3.
(a) XRD pattern of $Bi_{1-x}Pb_xFeO_3$ multiferroic [11]. (b) Glancing angle X-ray diffraction (GAXRD) pattern of $BiFeO_3/BaTiO_3$ [thickness of BTO = 100 nm; while BFO = 50 nm (BFBT-5), 100 nm (BFBT-10), 150 nm (BFBT-15), and 200 nm (BFBT-20)] [30]. (c) Raman spectra of $BiFeO_3$ [31].

3.1.2 Crystalline structure of $BiFeO_3/BaTiO_3$ bilayer interface

Glancing angle XRD patterns, recorded at incident angle 1°, on different BFO/BTO bilayer thin films sputtered on $Pt/TiO_2/SiO_2/Si(100)$ substrates shown in **Figure 3(b)** [30]. Both BFO and BTO layers were found to be polycrystalline in nature without any impurity phase. The (110) plane of BFO appeared in θ-2θ mode only which was not be observed in GAXRD mode. It can be inferred that the bottom BTO layer promotes the formation of pure perovskite phase and high degree of (110) orientation in film texture.

3.1.3 Raman spectra of $BiFeO_3$ nanoparticles

The Raman spectrum of BFO obtained using 488 nm excitation wavelength depicted in **Figure 3(c)** [31]. The spectra have been deconvoluted into 10 individual components for BFO (3 A modes and 7 E modes). It was studied by DFT calculation of first principle that the low frequency Raman modes below 167 cm^{-1} are due to Bi atoms, and the modes between 152 and 262 cm^{-1} are due to Fe atoms [31].

However, oxygen atoms dominated with higher frequency modes above 262 cm^{-1}. The shifting and broadening of Raman modes with standard values suggests the presence of disorder and oxygen vacancies, and the internal microstrain due to Fe ions which might to change resulting magnetism.

3.2 Microstructural studies of BiFeO$_3$

3.2.1 FESEM image of $Bi_{0.9}Pb_{0.1}FeO_3$

Figure 4(a) shows the FESEM image of $Bi_{0.9}Pb_{0.1}FeO_3$ nanostructure [11]. The diameter of the nanorods, D = 125 ± 4 nm, and length, L = 900 ± 20 nm. However for pure BFO, the nanoparticles (D = 75 ± 2 nm) are formed [11]. This morphological variation with Pb doping into BFO is explained due to variation in the valence states of Fe ions due to oxygen vacancies. The substitution of Pb^{2+} into Bi^{3+} ions induces Fe^{2+}/Fe^{3+} ions in the BFO matrix that can influence lattice defects (oxygen vacancies) in the rhombohedral BFO phase. It results into an anisotropic growth along the *c*-axis.

3.2.2 $BiFeO_3$-$CoFe_2O_4$ self-assembled nanocomposite

Figure 4(b) shows a top-view SEM image of a square array with period 83 nm [32]. The bright rectangular islands visible in this image correspond to (111)-faceted tops of the $CoFe_2O_4$ pillars, while the darker area corresponds to single crystal

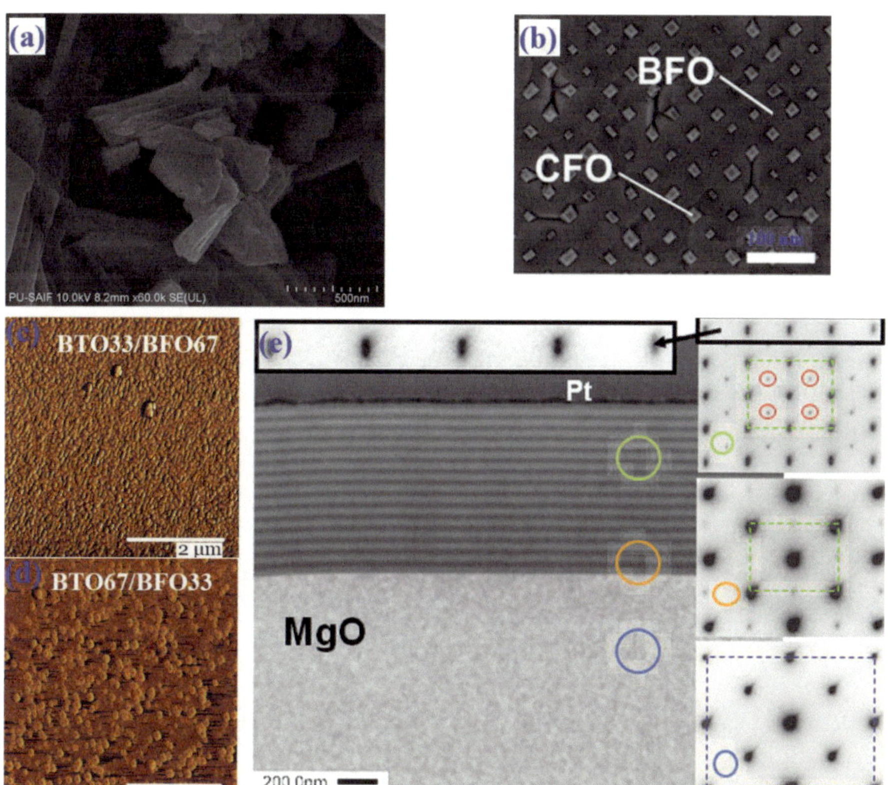

Figure 4.
(a) FESEM image of $Bi_{0.9}Pb_{0.1}FeO_3$ [11]. (b) Top-view SEM image of BFO-CFO nanocomposite [32]. (c and d) AFM surface images of BTO33/BFO67 and BTO67/BFO33 [33]. (e) STEM dark-field image of ($BaTiO_3$-$BiFeO_3$) × 15 grown on MgO [34].

$BiFeO_3$ matrix. This BFO-CFO nanocomposite was grown by pulsed laser deposition. The film thickness was between 50 and 100 nm.

3.2.3 Atomic force microscopy images of BTO/BFO thin films

The $BiFeO_3$ and $BaTiO_3$ were used to grow homogeneous composite thin films and multilayer heterostructures with 15 double layers by pulsed laser deposition [33]. The thin films are composites grown directly from mixed PLD targets with 67 wt% BTO/33 wt% BFO (BTO67/BFO33) and 33 wt% BTO/67 wt% BFO (BTO33/BFO67), respectively. **Figure 4(c)** and **(d)** are the AFM surface images of these two different composite films and the nanoparticle sizes are in the ranges 40-100 nm and 100-200 nm, respectively.

3.2.4 Scanning transmission electron microscopy (STEM) of $BaTiO_3$-$BiFeO_3$

Multiferroic ($BaTiO_3$-$BiFeO_3$) $\times$ 15 multilayer heterostructures show high ME coefficients, α_{ME} up to 24 V cm^{-1} Oe^{-1} at 300 K [34]. The STEM and SAED mapping results of the ($BaTiO_3$-$BiFeO_3$) $\times$ 15 multilayer are depicted in **Figure 4(e)**. The STEM cross section is shown 15 double layers $BaTiO_3$-$BiFeO_3$ at smooth interfaces. The octahedral tilt involves both clockwise and counter clockwise rotations around [111], which is parallel to ferroelectric dipole displacements to *R3m* phase. The difference between the lattice parameters of the tetragonal $BaTiO_3$ and rhombohedral $BiFeO_3$ layers can be detected by reflection splitting along the growth direction (inset in **Figure 4(e)** top).

3.3 Ferroelectric behavior of $BiFeO_3$

The $BiFeO_3$ is a rhombohedrally distorted perovskite material, which means that the ferroelectric polarization can have orientation along the four pseudo-cubic diagonals (<111>) [35–38]. The largest relative displacements are those of Bi relative to O, consistent with a stereochemically active Bi lone pair that might to induce ferroelectricity of $BiFeO_3$.

3.3.1 Ferroelectric polarization of Pb doped $BiFeO_3$ nanoparticles

The ferroelectric polarization of multiferroic $Bi_{1-x}Pb_xFeO_3$ nanostructures is given in **Table 2** [11]. In BFO, the lone-pair orbital of Bi^{3+} ($6s^2$) is stereochemically active and responsible for ferroelectric distortion. Here, the distortion is induced by Pb doping and therefore, by tuning the lone-pair activity. Generally, the ferroelectric behavior is weakened due to an increase in oxygen vacancies that form more

Sample	P_s ($\mu C\ cm^{-2}$)	P_r ($\mu C\ cm^{-2}$)	E_c ($kV\ cm^{-2}$)	J ($\mu A\ cm^{-2}$)	ε		MC (%)		T_{FE} (K)
					1 kHz	1 MHz	1 kHz	1 MHz	
BFO	0.75	0.35	7.66	10.31	6.2	5.1	0.68	0.61	644
BPFO5	1.76	0.63	6.03	6.27	1.9	1.8	2.56	1.59	631
BPFO75	3.55	1.57	8.55	2.72	13	10.6	3.61	0.36	649
BPFO10	6.73	2.63	5.06	1.57	34	12	0.86	0.11	629

Table 2.
Values of spontaneous polarization (P_s), remanent polarization (P_r), electric coercivity (E_c), current density (J) at 20 kV cm^{-1}, dielectric constant (ε) and MC at 1 kHz and 1 MHz, and ferroelectric phase transition (T_{FE}) at 1 MHz for $Bi_{1-x}Pb_xFeO_3$ [x = 0 (BFO), 0.05 (BPFO5), 0.075 (BPFO75) and 0.1 (BPFO10)] nanostructures [11].

free electrons. It resulting into a higher conductivity and hence have a harmful influence on the ferroelectricity. The Pb doping into BFO could increase the grain size as well the oriented growth (nanorod-type) in the samples. From **Table 2**, there is a considerable reduction in the leakage current of BFO upon Pb^{2+} doping, which indicates reduction in oxygen vacancies.

3.3.2 Ferroelectric and piezoelectric properties of epitaxial $BiFeO_3$ thin film

Figure 5(a) and **(b)** shows the ferroelectric polarization and piezoelectric behavior of epitaxial $BiFeO_3$ thin film [35]. The films display a room-temperature spontaneous polarization (50 to 60 μC cm^{-2}) almost an order of magnitude higher than that of the bulk (6.1 μC cm^{-2}). These results leads to the observations of

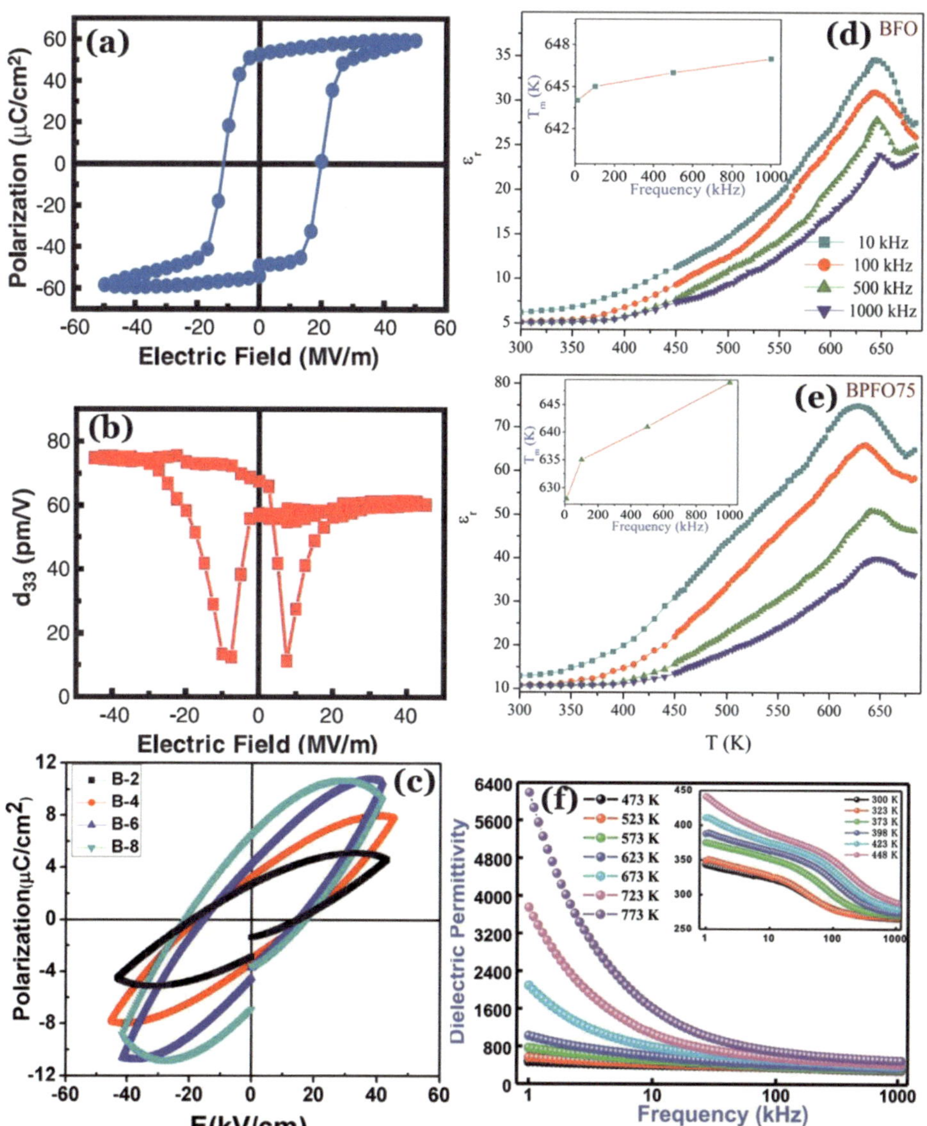

Figure 5.
(a) A ferroelectric hysteresis of $BiFeO_3$ thin film at 15 kHz. (b) a small signal d_{33} for a 50-μm capacitor [35]. (c) Room temperature PE loops of $BaTiO_3/BiFeO_3/BaTiO_3$ trilayer thin films having different $BiFeO_3$ thicknesses [36]. Temperature dependent relative permittivity (ε_r) of (d) $BiFeO_3$ (e) $Bi_{0.925}Pb_{0.075}FeO_3$ [11]. (f) Frequencies dependent dielectric permittivity at different temperatures for $Bi(Co_{0.4}Ti_{0.4}Fe_{0.2})O_3$ (insets shows lower temperature behavior) [37].

heteroepitaxial, in-plane compressive stress imposed by the epitaxial bottom electrode allows growth of a monoclinic crystal structure in BFO, and the degree of compressive stress progressively decreases with increasing BFO thickness. The piezoelectric hysteresis loop shows a remanent out-of-plane piezoelectric coefficient (d_{33}) value 70 pm V^{-1}, representing the piezoresponse of the film in the fully clamped state.

3.3.3 Ferroelectric hysteresis of $BaTiO_3/BiFeO_3/BaTiO_3$

Figure 5(c) represents polarization hysteresis loops of trilayer films of $BaTiO_3$/$BiFeO_3$/$BaTiO_3$ measured at 50 kV cm^{-1} applied electric field frequency of 10 kHz [36]. This trilayer thin film was prepared by RF-magnetron sputtering technique at different thicknesses of $BiFeO_3$ layer. The thickness of BTO layer is 20 nm at the top and bottom, and the middle layer BFO is deposited with thicknesses of 20 nm (B-2), 40 nm (B-4), 60 nm (B-6), and 80 nm (B-8), respectively. The film showed maximum remnant electric polarization ($2P_r$) of 13.5 μC cm^{-2} and saturation magnetization (M_s) of 61 emu cc^{-1} at room temperature. The ferroelectric polarization was found to be improved with increasing thickness of BFO layer may be attributed to the reduced oxygen vacancies.

3.4 Dielectric properties of $BiFeO_3$

3.4.1 Frequency dependent dielectric properties of $Bi(Co_{0.4}Ti_{0.4}Fe_{0.2})O_3$

Figure 5(f) correlates the dielectric permittivity versus frequency plot at temperatures from 300 to 773 K of $Bi(Co_{0.4}Ti_{0.4}Fe_{0.2})O_3$ multiferroic [37]. As the value of dielectric permittivity decreases upon increasing frequency, its nature could be described by Koop's hypothesis and Maxwell-Wagner mechanism. The nano grains with highly resistive grain boundaries might exist in an inhomogeneous medium from which the application of electric field constructs space charge polarization. With low frequency, the grain boundaries influence is more dominant to cause dispersion in dielectric properties. However, the higher frequency reduces the space charge polarization impact because the slow traveling species are not capable to trace an applied electric field.

3.4.2 Temperature dependent dielectric permittivity of Pb:$BiFeO_3$

Figure 5(d) and **(e)** shows the temperature dependent relative permittivity (ε_r) for $BiFeO_3$ and $Bi_{0.925}Pb_{0.075}FeO_3$ (BPFO75) multiferroics [11]. The value of ε_r starts to increase with temperature because of T_C for pure BFO is 1103 K. This change in ε_r at 600-650 K for both Pb:BFO samples occurs due to occurrence of T_N. For pure BFO, the value of T_N is 643 K. The value of the ferroelectric phase transition (T_{FE}) is 644 and 649, respectively, for BFO and BPFO75. This observation is an anomaly in the phase transition; T_{FE} around T_N confirms the ME coupling, which must be correlated with inverse DM-type interactions and Landau-Devonshire theory of phase transition. The variation in phase transition temperature with frequency for Pb:BFO nanostructures (inset of **Figure 5(d)** and **(e)**) indicates the emergence of the relaxor behavior which explained with an increase cation disorder in the B-site and Bi-site substitution by Pb^{2+}.

3.5 Magnetization in $BiFeO_3$

The magnetization of $BiFeO_3$ is reported in the refs. [39–44]. The cycloidal model of spin ordering in BFO is distorted at low temperatures. Any break to the cycloidal spin structure could induce uncompensated spins, enhancing the magnetization [45].

3.5.1 Magnetization in multiferroic $BiFeO_3$ and $Bi_{0.9}Pb_{0.1}FeO_3$

The ferromagnetic behavior of Pb substituted $BiFeO_3$ is reported in ref. [11] that the maximum value of magnetization, M = 4.73, 8.41, 2.62 and 8.99 emu g^{-1}, respectively, for BFO, BPFO5, BPFO75 and BPFO10. The origin of the variation of magnetization is analyzed with temperature dependent zero field (ZFC) and field cooled (FC) magnetization is shown in **Figure 6(a)**. The splitting of the ZFC/FC curves usually appears as the co-existent system of the antiferromagnetic and ferromagnetic phases. With nanostructural dimensions of less than 62 nm, there is a possibility of modification to the cycloidal spin structure of BFO, and that can lead to weak room temperature ferromagnetism. The sharp cusp observed around 65 and 79 K, respectively, for BFO and BPFO10 nanostructures in the ZFC curve is represented by the blocking temperature (T_B which may attribute via superparamagnetic relaxation, glass transition, T_N for antiferromagnetic-

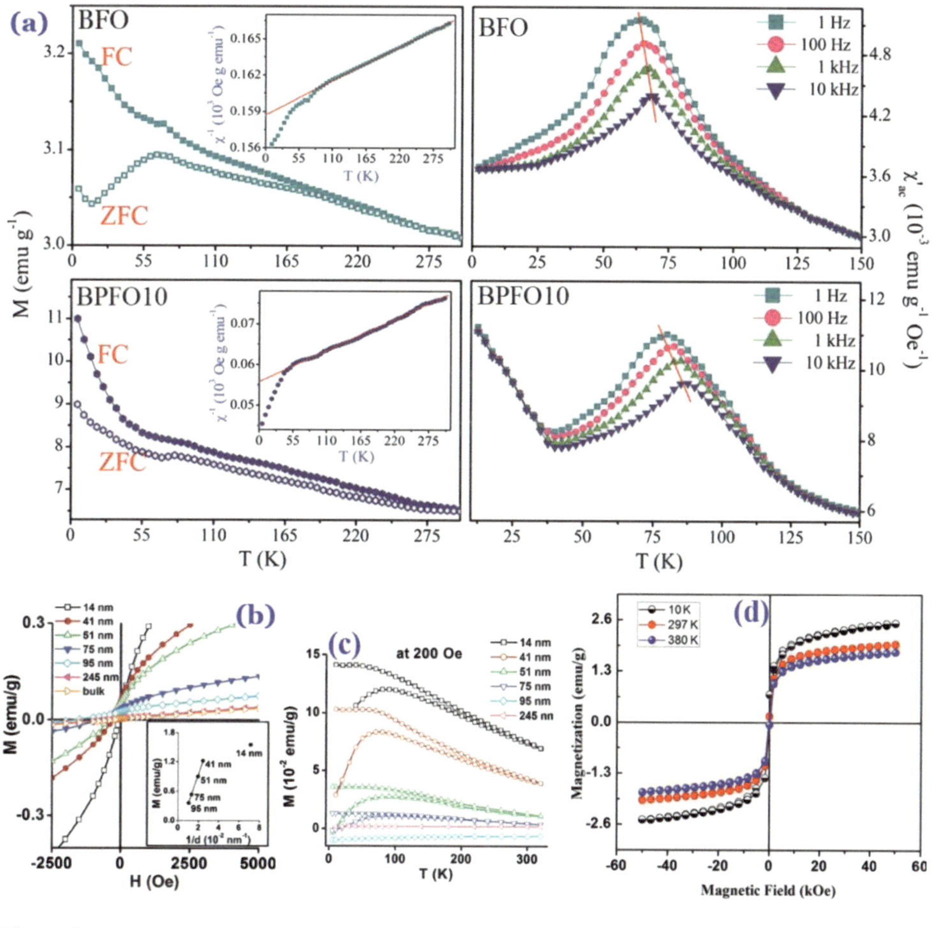

Figure 6.
(a) Magnetization (M) as a function of temperature (T) following ZFC and FC at H = 500 Oe for $BiFeO_3$ (BFO) and $Bi_{0.9}Pb_{0.1}FeO_3$ (BPFO10) multiferroics. The respective insets are $\chi^{-1}(T)$ following Curie-Weiss law. (a: right) The temperature dependent real part of the ac *magnetic susceptibility (χ'_{ac}) at T = 5-200 K (H_{ac} = 2.5 Oe without any dc field bias) [11]. (b) Magnetic hysteresis at 300 K for $BiFeO_3$ nanoparticles with different nano-sizes. Inset shows the corresponding magnetization at 50 kOe as a function of 1/d. (c) Respective ZFC and FC curves at 200 Oe [39]. (d) Magnetic hysteresis of $BiFeO_3$ nanoparticles grafted on graphene nanosheets ($BiFeO_3$-g-GNS) [40].*

ferromagnetic transition). The enhancement in magnetization is explained by Coey JMD *et al.* [46] model of F-center exchange mechanism where spin-polarized electrons were trapped at oxygen vacancies to cause higher magnetic moments. The upward curvature in FC curve of M(T) measurements of Pb:BFO suggested a Curie-Weiss like behavior **Figure 6(a)** (inset). The estimated value of θ is found to be negative which indicate to the formation of antiferromagnetic interactions.

The *ac* magnetic susceptibility of Pb substituted BFO is measured at 1 Hz, 100 Hz, 1 kHz 10 kHz, and the temperature dependent real (χ'_{ac}) *ac* magnetic susceptibility (χ_{ac}) is shown in **Figure 6(a)**(right). The applied oscillating field, H_{ac} = 2.5 Oe without any *dc* bias in T = 5-200 K. A quite sharp cusp is observed in both the samples. This *ac* magnetic measurement at different frequencies revealed the peak positions of χ'_{ac}(T) curve shift toward higher temperature and the peak magnitudes drop down with rising frequency. Such behavior is expected for a spin glass system. The dynamic susceptibility measurements can thus be used to confirm such spin glass or superparamagnetic by using frequency dependence of $T_f(\omega)$ in the expression, $\Delta p = \frac{\Delta T_f}{T_f \Delta(\log \omega)}$. The calculated peak shift (Δp) per decade of frequency shift has a value 0.014 and 0.019, respectively, for BFO and BPFO10. These values of Δp are lower than those observed for superparamagnetic system (Δp is ~0.1.54).

3.5.2 Nano size dependent magnetism of $BiFeO_3$

The SQUID results as shown in **Figure 6(b)** suggest that a magnetic response in $BiFeO_3$ can be initiated when the size of the system is less than about 95 nm [39]. A plot of the magnetization, measured at the maximum applied field of H_{appl} ~ 50 kOe as a function of 1/d, is given (inset of **Figure 6(b)**). Neel L [47] attributed the magnetic moment of small antiferromagnetic particles for the incomplete magnetic compensation between these two spin sublattices. For single-domain antiferromagnetic particles, the magnetization is expected to scale as ~1/d(diameter), that is, as the surface to volume ratio [39]. For particles ranging in diameter from 95 to 41 nm, a linear dependence is observed, indicating that the simple Neel model is applicable [48]. The smallest nanoparticle is 14 nm that deviates from the expected behavior which indicates that such 14 nm nanoparticle may diminish the model for superposition of an antiferromagnetic core and a ferromagnetic surface. The maximum magnetization, obtained as M_s ~ 1.55 emu g^{-1} for the 14 nm particles. **Figure 6(c)** shows the magnetization measurements as a function of temperature at an applied field strength of 200 Oe following ZFC and FC process. It is noted that the apparent sharp cusps observed in the magnetization curves at 50 K are reproducible for BFO samples with particle dimensions over 95 nm (e.g. 245 nm and bulk). For $BiFeO_3$ nanoparticles possessing diameters of ≤95 nm, associated data curves exhibit a broad magnetization maximum around T_{max} = 85 K. T_{max} represents a spin-glass-like freezing temperature due to high packing volume fraction as well as a complex interplay between finite size effects, interparticle interactions, and a random distribution of anisotropy axes.

3.5.3 Temperature dependent magnetization of $BiFeO_3$-g-GNS nanoparticles

Magnetic properties of $BiFeO_3$ grafted on graphene nanosheets ($BiFeO_3$-*g*-GNS) is studied using SQUID VSM with an applied field of 5 T at different 10, 297 and 380 K **Figure 6(d)** [40]. Magnetic moments of nanoparticles get distorted at higher temperature. The maximum observed value of magnetization is 2.52 emu g^{-1} at 10 K, while the minimum is 1.78 emu g^{-1} at 380 K. At room temperature, magnetization measured to be 1.98 emu g^{-1}. These values well matched with reported data [49]. It means the magnetic properties of $BiFeO_3$ are not compromised on GNS grafting.

3.5.4 Magnetization of $BiFeO_3$-$CoFe_2O_4$ (BFO-CFO) bulk heterojunction

Figure 7(a) exhibits the magnetic hysteresis in isotropic magnetic behavior along in plane and out of plane directions [41]. The value of M_s of BFO-CFO/mica is ~237 emu cm^{-3} with $H_c \sim 2$ kOe, which is smaller than from epitaxial CFO/STO (~3 kOe). This may due to the effect of an effective relaxation of clamping from the mica substrate.

3.5.5 Ferromagnetism in $BiFeO_3/BaTiO_3$ bilayer interface

Figure 7(b) shows the ferromagnetic behavior of $BiFeO_3/BaTiO_3$ (BTO thickness = 100 nm; BFO = 50 nm (BFBT-5), 100 nm (BFBT-10), 150 nm (BFBT-15), and 200 nm (BFBT-20) films [30]. The observed ferromagnetism in the bilayer thin films can be interpreted due to creation of unbalanced spins at the interface. Maximum magnetization value $M_s \sim 33$ emu cc^{-1} was observed in BFBT-5. The value of saturation magnetization is 20 emu cc^{-1} for BFBT-10 sample which is higher than for those observed in BFBT-15 (15 emu cc^{-1}) and BFBT-20 (8 emu cc^{-1}). This is because with smaller antiferromagnetic nanoparticles, the size reduction has incomplete surface compensation of long-range antiferromagnetic ordering which result into increase magnetic moment at comparatively smaller size nanograins.

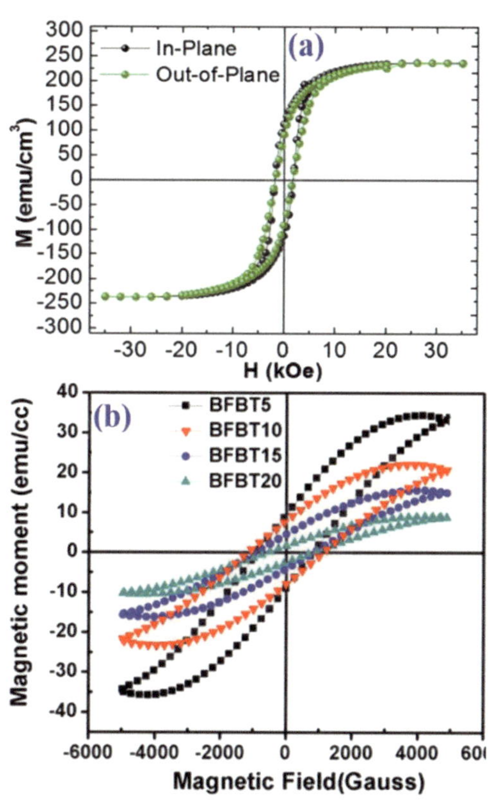

Figure 7.
(a) M-H hysteresis for $BiFeO_3$-$CoFe_2O_4$ composite at room temperature [41]. (b) M-H loops of bilayer $BiFeO_3/BaTiO_3$ thin films [30].

3.6 Magnetocapacitance

The ME behavior due to Magnetocapacitance (MC) effect in $BiFeO_3$ is given in the refs. [50–54]. The MC effect is the change in the capacitance with an external applied magnetic field. This MC/magnetodielectric effect has a resistive origin that arises from the Maxwell-Wagner effect and magnetoresistance [55].

3.6.1 Dielectric constant and MC of Pb substituted $BiFeO_3$

Table 2 shows the magnetic field affected dielectric constant of Pb:BFO nanostructures measured at room temperature. The frequency dependent relative permittivity (ε_r) of Pb:BFO in the frequency region 20 Hz-10 MHz under *dc* magnetic field (H = 0, 1 kOe) is given [11]. From **Table 2**, the improvement in dielectric constant with the substitution of Pb^{2+} for Bi^{3+} provides a larger vibration space to a larger dipole moment. Besides the oxygen vacancies due to ionic formation of Fe^{2+}/Fe^{3+} valence states, the shape/size of BFO grains might to influence the dielectric behaviors [11]. It is also observed from **Table 2** that the capacitance varies with applying magnetic field of 1 kOe, which indicate a positive/negative MC effect. The applied magnetic field leads to local stresses (or strains) and consequently changes in the polarization of the ferroelectric phase due to the piezoelectric effect. The values of MC $\{[\varepsilon(H) - \varepsilon(0) = \Delta\varepsilon]/\varepsilon(0)\}$ at a frequency of 1 MHz is 0.61, 1.59, 0.36 and 0.11%, respectively, calculated for BFO, BPFO5, BPFO75 and BPFO10 multiferroic.

3.6.2 Magnetocapacitance effect in $BiFe_{0.95}Sc_{0.05}O_3$

The magnetic field dependent capacitance for $BiFe_{0.95}Sc_{0.05}O_3$ system to induce MC effect at 30 kHz is shown in **Figure 8(a)** [50]. The value of MC is 0.04% at applied magnetic field of 5 T which is higher than from pure BFO (0.007%). This type of MC behavior might be correlated with P^2M^2 in a Ginzburg-Landau free energy leads to a quadratic dependent dielectric constant in respect to magnetization. The observed results of MC in **Figure 8(a)** have magnetization-linear dielectric behavior which may proportional to P^2M of a linear MC effect.

3.6.3 Piezoelectric properties of $BiFeO_3/Na_{0.5}Bi_{4.5}Ti_4O_{15}$ composite films

Lead-free $BiFeO_3/Na_{0.5}Bi_{4.5}Ti_4O_{15}$ (BFO/NBTO) composite films were deposited on Pt(100)/Ti/SiO_2/Si substrates using chemical solution deposition [51]. A giant ME voltage coefficient has maximum α_E = 136 mV cm^{-1} Oe^{-1} at H_{bias} = 8.0 kOe. **Figure 8(b)** shows the piezoelectric coefficient (d_{33}) and surface displacement (d) vs. applied voltage (V) for BFO/NBTO films. A typical butterfly curve from D-V characteristics shows maximum value 4.06 nm at 15 V of 2.63% highest ratio of strain. The occurrence of d_{33}-V piezoelectric hysteresis from D-V curve is the result converse piezoelectric effect. The d_{33}-V loop clearly shows that BFO/NBTO composite films are switchable and the ferroelectricity is retained. The piezoelectric coefficient d_{33} of BFO/NBTO films is as high as 285 pm V^{-1} at 20 V, which suggests the strong piezoelectric effect for BFO/NBTO films.

3.6.4 Magnetostriction of $BiFeO_3$-$BaTiO_3$, $Bi_{0.8}FeO_{2.7}$-$BaTiO_3$, and $BaFe_{12}O_{19}$

The magnetostriction of BFO-BTO, $B_{0.8}FO_{2.7}$-BTO, and $BaFe_{12}O_{19}$ ceramics was measured using a resistive strain gauge when a dc magnetic field was applied to the materials (**Figure 8(c)**) [52]. A contraction induced by the magnetic field is

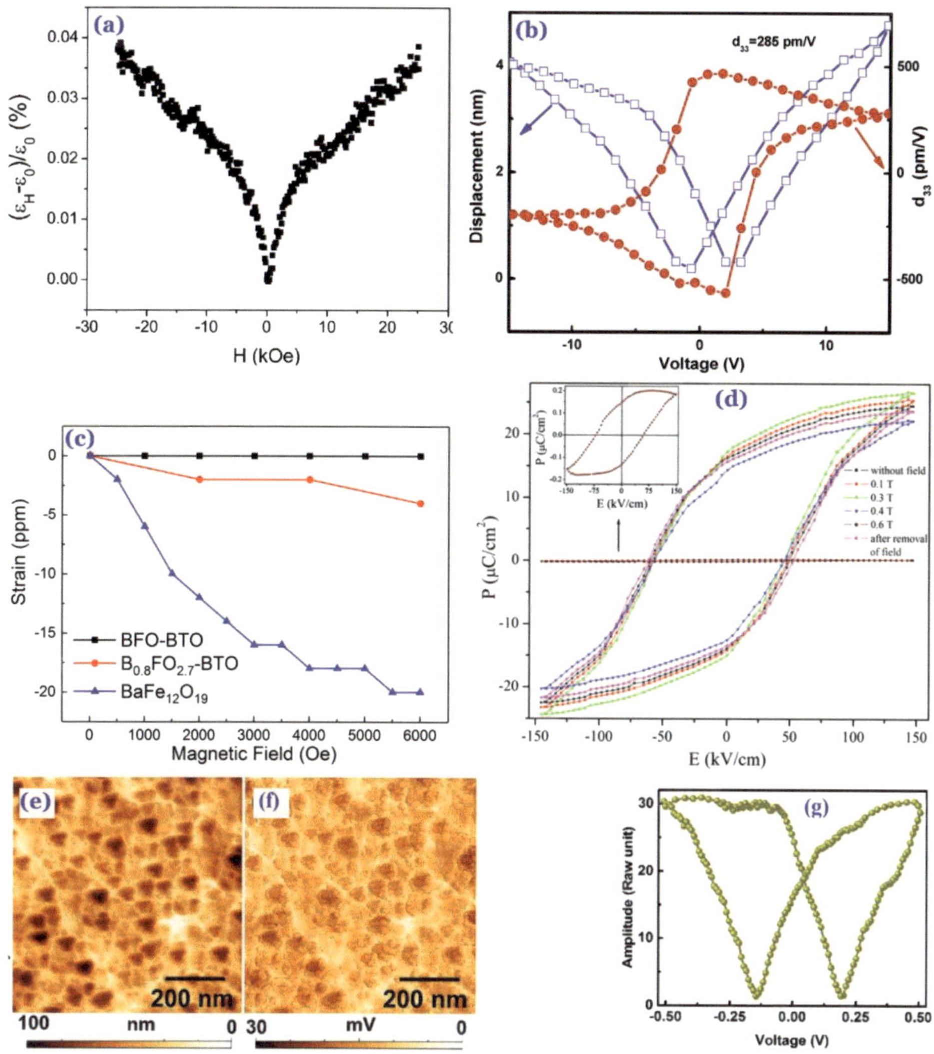

Figure 8.
(a) Magnetocapacitance of $BiFe_{0.95}Sc_{0.05}O_3$ [50]. (b) Piezoelectric response of $BiFeO_3/Na_{0.5}Bi_{4.5}Ti_4O_{15}$ (BFO/NBTO) composite [51]. (c) Magnetostriction of $BiFeO_3$-$BaTiO_3$, $Bi_{0.8}FeO_{2.7}$-$BaTiO_3$, and $BaFe_{12}O_{19}$ measured at room temperature [52]. (d) P-E hysteresis under 0-0.6 T field for BFO-BTO nanoparticles. (Inset of d) P ~ 0 at 0.6 T [53]. PFM study on $BiFeO_3$ thin film: (e) AFM (f) amplitude image (g) amplitude behavior at an applied bias [54].

observed, and at 6 kOe, the contraction is approximately 20 ppm. $BaFe_{12}O_{19}$ is a ferromagnetic material and can have a magnetostrictive response. However, due to the weak ferromagnetic nature of the BFO-BTO ceramics, it was difficult to detect the strain below 6 kOe. Because only a small amount of $BaFe_{12}O_{19}$ is generated in the Bi-deficient ceramics, the magnetostriction is also very small in the materials.

3.6.5 ME coupling: ferroelectric polarization in an applied magnetic field

The mechanism for the spin driven ferroelectricity must be involved the spin-current model or inverse DM interaction due to the local electric polarization, $p \propto e_{ij} \times (S_i \times S_j)$, where e_{ij} is the unit vector of adjacent spins, S_i and S_j [56]. These cycloidal spin structures produced the macroscopic electric polarization P due to helicity of spins. The ferroelectric hysteresis under an external magnetic field and the flop near 0.6 T reveal a strong ME coupling in BFO-BTO multiferroics as shown in **Figure 8(d)** [53]. Without applying H, values of P_{max} = 24.80 μC cm^{-2}, P_r = 15.13 μC cm^{-2} and E_c = 53.6 kV cm^{-1} are observed. Switching from +P_r to -P_r by E, and magnetic switching from +P_r to zero at 0.6 T have been observed. The inset of **Figure 8(d)** shows a low polarization response with lossy hysteresis of BFO-BTO composite at 0.6 T, which may cause by the electrode. When a magnetic field is applied to a ME material, the material is under strain to induce a stress on the piezoelectrics (ferroelectric) to orient ferroelectric domains, leading to enhance polarization.

3.6.6 ME effect in $BiFeO_3$ by PFM

Figure 8(e) shows AFM, out of plane piezoresponse amplitude of the BFO thin film which leads to agglomeration of the nanoparticles of average size 10 nm [55]. It can be seen from **Figure 8(f)** that BFO nanoparticles exhibits positive and negative polarization components and the ferroelectric domains are constrained at grain boundaries. The PFM amplitude under bias voltage is shown in **Figure 8(g)**. It observed ferroelectric hysteresis for dc voltage sweeps in +0.5 V to −0.5 V to +0.5 V which indicates that sub - 5 nm BFO nanoparticles retains their ferroelectric behavior that might be usable for read-write operation.

3.7 Magnetoelectric (ME) coupling

3.7.1 ME coupling in $Bi_{0.88}Dy_{0.12}Fe_{0.97}Ti_{0.03}O_{3+\delta}$ (BDFO) and $BiFeO_3$

The ME effect of BFO and BDFO was measured, and the results are shown in **Figure 9(a)** [57]. It can be seen that the pure $BiFeO_3$ exhibits no ME signal under bias magnetic field because spatially incommensurately modulated spin structure that cancels out the linear ME effect. In contrast, BDFO exhibits a strong ME signal under bias magnetic field at 300 K due to the ME coefficient α_E-magnetic field (α_E-H) hysteresis to include the features of saturation at field of 250 Oe. This is because the switching of electric polarization by either 109° or 71° leads to switching of the ferroelastic domain states [58–60]. The incorporation of Dy^{3+} into BFO suppresses the spiral spin structure, leads to weak ferromagnetism. Because the magnetic moment lags behind the variation of magnetic field, the electric polarization induced by magnetic field through ME effect is also lagged behind the variation of magnetic field, giving α_E(H) hysteresis.

3.7.2 ME coupling of 15 × (10 nm $BaTiO_3$-5 nm $BiFeO_3$)

The ME coefficient was measured as a function of dc bias field in **Figure 9(b)** for 15 × (10 nm $BaTiO_3$-5 nm $BiFeO_3$) [58]. The ME coefficient reaches its maximum, off - resonance, value of 60.2 V cm^{-1} Oe^{-1} at a bias field of 2 T. This higher value is described due to the rotation of the oxygen octahedra with the antiferrodistortion vector Ω and shows that the linear ME effect depends on the derivative of Ω with respect to the external field. This leads to giant values of α_{ME}.

3.7.3 ME coupling of 15 × [$BaTiO_3$/$BiFeO_3$] multilayer

The effect of strain and interfaces on the ME coupling in $BiFeO_3$ was studied for a number of thin films and multilayers of 15 × [$BaTiO_3$/$BiFeO_3$] as shown in **Figure 9(c)** [33, 59]. To this end, a direct longitudinal ac method was used to measure the ME coefficient α_{ME} as a function of static magnetic field. It is clear that α_{ME} of the multilayer is notably larger than that of the BFO film. Since the

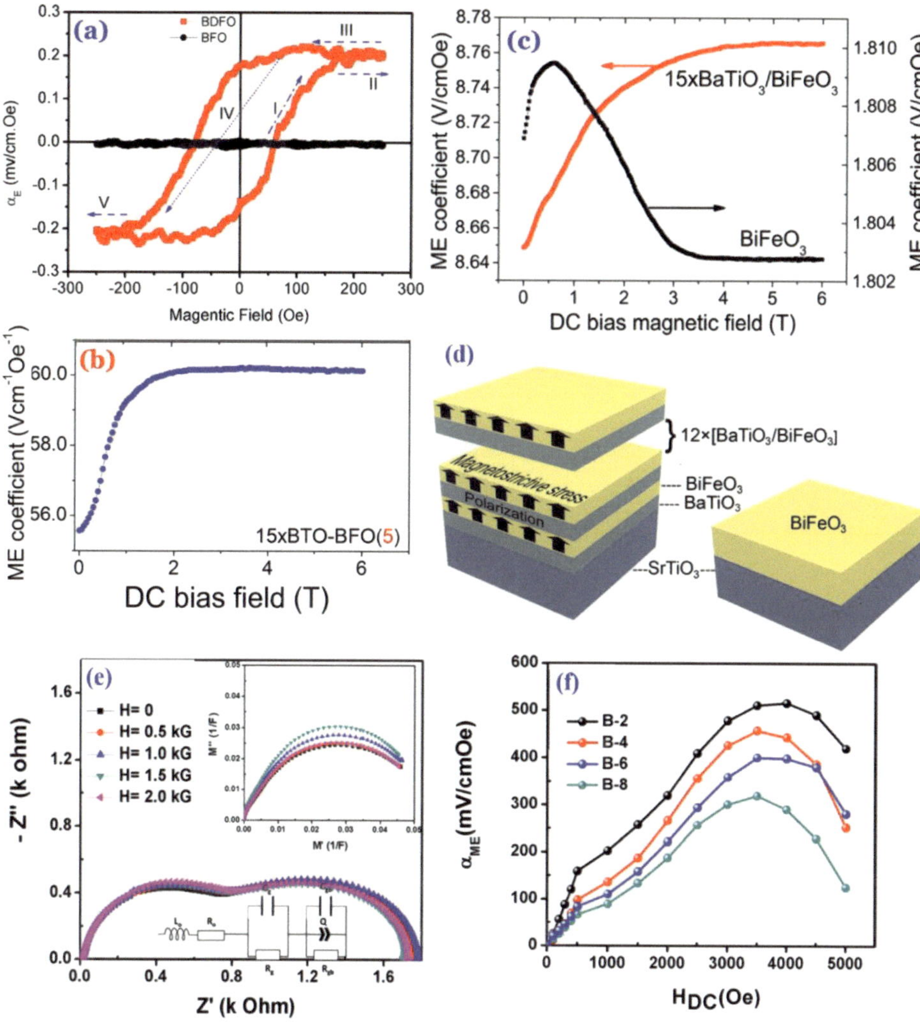

Figure 9.
(a) Magnetoelectric P(H) hysteresis at 300 K, showing ME coefficient of $Bi_{0.88}Dy_{0.12}Fe_{0.97}Ti_{0.03}O_{3+\delta}$ (BDFO) and BFO multiferroic [57]. (b) ME coefficient of the multilayers 15 × (10 nm $BaTiO_3$ - 5 nm $BiFeO_3$) at room temperature [58]. (c) ME coefficient α_{ME} of 15 × [$BaTiO_3$/$BiFeO_3$] multilayer film and pure $BiFeO_3$. (d) Ferroelectric-multiferroic multilayer (left) showing the ME effect through magnetostrictive-piezoelectric interface coupling (arrows represent magnetostrictive stress); and a $BiFeO_3$ film (right) [59]. (e) ME-impedance response of $BaTiO_3$/$BiFeO_3$/$BaTiO_3$ trilayer film measured at room temperature (magnetomodulus plots: inset). (f) ME coefficient for BTO/BFO/BTO trilayer films [36].

multilayer additionally contains piezoelectric/piezomagnetic interfaces, an extra ME coupling in multilayers may occur via the horizontal interfaces through strain-mediated interface coupling. The magnetostrictive stress is produced in the weak ferromagnetic $BiFeO_3$ layer and is transferred to ferroelectric $BaTiO_3$ layer through the interface (**Figure 9(d)**). This mechanical stress generates an electric potential difference in the ferroelectric layer via a piezoelectric effect.

3.7.4 Magnetoimpedance and ME effects of $BaTiO_3$/$BiFeO_3$/$BaTiO_3$ heterostructure

The ME effect for $BaTiO_3$/$BiFeO_3$/$BaTiO_3$ heterostructure is also investigated by analyzing complex impedance ($-Z''$ vs. Z') as well as complex modulus plots (M'' vs. M') under applied magnetic fields (**Figure 9(e)**) [36]. The data were fitted with an equivalent circuit of two series RC-elements. Two well resolved semicircles

$BiFeO_3$ composition	Synthesis method	Phase structure	Nano-structure	P_s/P_r ($\mu C\ cm^{-2}$)	M ($emu\ g^{-1}$)	MC (%)	α_{ME} $mVcm^{-1}\ Oe^{-1}$	Ref.
$BiFeO_3$	Pulsed laser deposition	M	70 nm thick film	60/55	150	—	3000	[35]
$BiFeO_3$	Chemical-solution deposition	R	140 nm thick film	125/73	—	—	—	[61]
$BiFeO_3$	Hydrothermal synthesis	R	20 nm square nanosheets	~4.2/3	3.1	—	—	[62]
$BiFeO_3$	Chemical combustion	R	D = 75 nm	0.75/0.35	0.73	0.68	—	[11]
$BiFeO_3$	PVA sol-gel	R	—	2.9/2.13	0.624	0.6	3.37	[63]
$Bi_{0.9}Ba_{0.1}Fe_{0.9}Mn_{0.1}O_3$	Mechano-synthesis	R	D = 1 μm	~1.1/0.82	AF	—	~74	[64]
$Bi_{0.9}Eu_{0.1}FeO_3$	Sol–gel	Distorted R	D = 27 nm	10.5/5.1	~0.7	11.9	—	[65]
$Bi_{0.9}Sm_{0.1}FeO_3$	Sol–gel	Distorted R	D = 28.5 nm	15.1/7.2	AF	~17	—	[66]
$BiFe_{0.925}Sc_{0.05}Ti_{0.025}O_3$	Sonochemical	Distorted R	D = 20-25 nm	2.63/1.03	0.34	0.05		[67]
$Bi_{0.9}Sm_{0.1}Fe_{0.95}Co_{0.05}O_3$	Sol–gel/pulsed laser deposition	R	300 nm thick film	22.91/16.1	35	—	142	[68]
$BiFe_{0.92}Mn_{0.08}O_3$	Solution-gelation technique (thin film)	R	D = 23.96 nm	23.56/9.5	—	—	—	[69]
$Bi_{0.9}Y_{0.1}FeO_3$	Spin coating deposition	R	D = 90-105 nm	—	2.87	—	—	[70]
$BiFe_{0.9}Co_{0.09}Mn_{0.01}O_3$	Hydrothermal route	R	—	—	10.88	—	7.4	[71]
$0.33Ba_{0.7}Ca_{0.3}TiO_3$-$0.67BiFeO_3$	Conventional sintering method	T/R	D = 12 μm	15/9.1	0.33	2.96	—	[72]
$(Bi_{1/2}Ba_{1/4}Sr_{1/4})(Fe_{1/2}Ti_{1/2})O_3$	Thermo-mechanical	R	D = 5-20 nm	~12/0.6	~0.05	—	8.7	[73]
$BaTiO_3$–$BiFeO_3$–$LaFeO_3$	Sol–gel	O	D = 1-2 μm	~2.5/0.14	0.67	—	354	[74]
5% $BiFeO_3$–$NaNbO_3$ in P (VDF-TrFE)	Spin coating	α & β phases of PVDF	D = 100-150 nm	8/2	0.03	—	2400	[75]
$0.75(Bi_{0.99}La_{0.01})FeO_3$-$0.25BaTiO_3$	Conventional solid state	R/C	Core-shell	~35/27	~0.6	—	~3.4	[76]
$0.9BiFeO_3$-$0.1BaTiO_3$	Sol–gel	R	D = 186 nm	~0.15/0.1	AF	—	2.74	[77]

$BiFeO_3$ composition	Synthesis method	Phase structure	Nano-structure	P_s/P_r ($\mu C\ cm^{-2}$)	M ($emu\ g^{-1}$)	MC (%)	α_{ME} $mVcm^{-1}\ Oe^{-1}$	Ref.
$0.5BiFeO_3/0.5MnFe_2O_4$	Sol–gel auto-combustion	R/spinel	D = 110 nm	0.64/0.97	15.65	—	—	[78]
$BiFeO_3/BiMnO_3$	Pulsed laser	Pseudo-cubic	Superlattices: ~32 nm thick	26	10	—	—	[79]
$BiFeO_3-BaTiO_3$	RF-magnetron sputtering	(110) plane: high degree	50 nm thick $BiFeO_3$	15.5/9	33	4.96	61	[30]
$0.7BiFeO_3/$ $0.3MgLa_{0.025}Fe_{1.975}O_4$	Sol–gel auto combustion	distorted R	D = 56 nm	0.28/0.08	5.28	—	—	[80]
$0.9BiFeO_3$-$0.1PbTiO_3$	Sol–gel solid state	R/T	—	3.8×10^{-5}	0.06	—	0.2	[81]
$BaFe_{12}O_{19}/BiFeO_3$	Mechano-chemical	H/R	D = 588 nm	~2.8/1.4	~28	—	11.9	[82]
$BiFeO_3/CoFe_2O_4$	Solvothermal	R/spinel	D = 285 nm	2.1/1	30.1	—	8.5	[83]
$0.7BiFeO_3$-$0.3PbTiO_3$	Solid-state	R/T	D = 37 nm	~1.1/0.2	AF	—	—	[84]
$0.67BaTiO_3$–$0.33BiFeO_3$	Pulsed laser	R/T	1850 nm thick film, D = 40 nm	$400/230 \times 10^{-2}$ Cm^{-2}	$2 \times 10^{-3} \mu_o T$	—	20.75	[33]
$BaTiO_3/BiFeO_3/BaTiO_3$	RF-magnetron sputtering	R/T	BTO = 20 nm BFO = 40 thick thin film	~8/3.2	37	5.95	457	[36]
$Bi_{0.8}FeO_{2.7}$-$BaTiO_3$	Solid-state	R/T	D = ~1 μm	~48/30.2	~5.5	—	~120	[52]
$BiFeO_3/Na_{0.5}Bi_{4.5}Ti_4O_{15}$	Chemical solution deposition	R/O	BFO = 879 nm NBT = 545 nm	62.1/38.7	4.68	—	136	[51]

Table 3.
Summary of multiferroic properties of $BiFeO_3$ [phase structure (monoclinic (M), rhombohedral (R), tetragonal (T), orthorhombic (O), cubic (C), hexagonal (H)), nanostructural size (diameter (D) and length (l)), ferroelectric spontaneous polarization (P_s), remanent polarization (P_r), saturation magnetization (M), antiferromagnetic (AF), magnetocapacitance (MC), and magnetoelectric coefficient (α_{ME})].

representing increase in both grain and grain boundary resistance (R_g and R_{gb}) with applied magnetic field are shown. A maximum 20% increase in grain capacitance (C_g) with applied magnetic field of 2 kG to represent an intrinsic ME effect. The bonding between Fe and Ti atoms at interface results into ME interaction between BFO and BTO at both interfaces. This interaction to change grains/boundaries resistances with the application of magnetic field induced magnetoimpedance/MC effect which is explained with Maxwell-Wagner model consisting of two leaky capacitors connected in series.

The ME coefficient, α_{ME} was measured in trilayer $BaTiO_3/BiFeO_3/BaTiO_3$ film by dynamic method (**Figure 9(f)**) and the detailed measurement set-up is given [36]. The α_{ME} was calculated using equation, $\alpha_{ME} = \delta V/\delta H.t$, where δV is the measured output voltage, δH is applied ac magnetic field, and *t* is the film thickness. The maximum α_{ME} of ~515 mV cm^{-1} Oe^{-1} is observed for B-2 film. By increasing the thickness of BFO layer, α_{ME} found to be reduced to 457 mV cm^{-1} Oe^{-1} for B-4, 400 mV cm^{-1} Oe^{-1} for B-6, and to 318 mV cm^{-1} Oe^{-1} for B-8. The enhancement in ME coupling for trilayer films may the effect of bonding between Fe and Ti atoms at both interfaces via oxygen atom. The reduction in oxygen vacancies with increasing thickness of BFO layer results into decreasing α_{ME} value.

3.8 Comparison of multiferroic properties of $BiFeO_3$

In **Table 3**, we have reported the list of multiferroic properties such as synthesis method, phase structure, nanostructures, ferroelectric behavior, magnetization, magnetocapacitance and ME coefficient of $BiFeO_3$. It is observed that the single phase $BiFeO_3$ has multiferroic behavior enhanced due to different doping from transition and rare earth ions. For the composites of $BiFeO_3$, there are moderate improvements in ME coupling. However, for multilayer $BiFeO_3$ with $BaTiO_3$ or ferrites has remarkable value of ME coupling. The Magnetocapacitance effect study on $BiFeO_3$ is hardly reported. Therefore it is summarized that the different multi-layers perovskites structures of $BiFeO_3$ may give much advancement to the multiferroic behaviors.

4. Conclusion

Multiferroic $BiFeO_3$ becomes a suitable material for spintronic application of data storage. Wet chemical methods, hydrothermal, Polymer-directed solvothermal, sol–gel template process, sonochemical, anodized alumina template, sol–gel based electrospinning and microwave synthesis are the best synthesis routes to control the shape and size of $BiFeO_3$ nanostructures. These nanostructural shape and size of $BiFeO_3$ has much impact to control the magnetism and leakage current of $BiFeO_3$. In addition to change dopant level and composites with other materials (such as ferrites and other perovskites like $BaTiO_3$), the $BiFeO_3$ thin films especially multilayers gives remarkable results of ferroelectric polarization and ME voltage.

Acknowledgements

The author K.C. Verma thankfully acknowledges the financial support by UGC of Dr. DS Kothari Post Doctorate Fellowship [No. F4-2/2006(BSR)/PH/16-17/0066] and CSIR-HRDG for SRA (Pool Scientist) fellowship Grant No. B-12287 [SRA (Pool No): 9048-A].

Author details

Kuldeep Chand Verma[1,2]

1 CSIR-Central Scientific Instruments Organisation, Chandigarh 160030, India

2 Department of Physics, Panjab University, Chandigarh 160014, India

*Address all correspondence to: dkuldeep.physics@gmail.com;
kcv0309@gmail.com

References

[1] Baibich MN, Broto JM, Fert A, Nguyen F, Dau V, Petroff F, Etienne P, Creuzet G, Friederich A, Chazelas J. Giant magnetoresistance of (001)Fe/(001)Cr magnetic superlattices. Phys. Rev. Lett. 1988; **61**, 2472-2475.

[2] Verma KC, Kotnala RK, Goyal N. Multi-Functionality of Spintronic Materials: Nanoelectronics. Elsevier; 2019; **eBook ISBN:** 9780128133545, 153-215.

[3] Sharma P, Zhang Q, Sando D, Lei CH, Liu Y, Li J, Nagarajan V, Seidel J, Sci. Adv. 2017; **3**: e1700512 (1-8)

[4] Tsymbal EY. Spintronics: Electric toggling of magnets, Nature Mater. 2012; **11**, 12-13

[5] Bibes M, Barthelemy A, Multiferroics Towards a magnetoelectric memory. Nature Mater. 2008; 7, 425-426

[6] Scott JF. Data Storage Multiferroic memories. Nature Mater. 2007; **6**, 256-257

[7] Kent AD, Worledge DC. A new spin on magnetic memories. Nature Nanotechnology. 2015; **10**, 187-191

[8] Verma KC, Kotnala RK. Lattice Defects Induce Multiferroic Responses in Ce, La-Substituted $BaFe_{0.01}Ti_{0.99}O_3$ Nanostructures. J. Am. Ceram. Soc. 2016; **99[5]**, 1601-1608 (2016)

[9] Verma KC, Kumar M, Kotnala RK. Magnetoelectric, Raman, and XPS Properties of $Pb_{0.7}Sr_{0.3}[(Fe_{2/3}Ce_{1/3})_{0.012}Ti_{0.988}]O_3$ and $Pb_{0.7}Sr_{0.3}[(Fe_{2/3}La_{1/3})_{0.012}Ti_{0.988}]O_3$ nanoparticles. Metall. Mater. Trans. A. 2014; **45A**, 1409-1414

[10] Gajek M, Bibes M, Fusil S, Bouzehouane K, Fontcuberta J, Barthelemy A, Fert A. Tunnel junctions with multiferroic barriers. Nature mater. 2007; **6**, 296-302

[11] Verma KC, Kotnala RK. Tailoring the multiferroic behavior in $BiFeO_3$ nanostructures by Pb doping. RSC Adv., 2016; **6**, 57727-57738

[12] Fiebig M, Lottermoser T, Meier D, Trassin M. The evolution of multiferroics. Nature Rev.: Mater. 2016; **1**, 1-14

[13] Gross I, Akhtar W, Garcia V, Martinez LJ, Chouaieb S, Garcia K, Carretero C, Barthelemy A, Appel P, Maletinsky P, Kim JV, Chauleau JY, Jaouen N, Viret M, Bibes M, Fusil S, Jacques V. Real-space imaging of non-collinear antiferromagnetic order with a single-spin magnetometer. Nature 2017; **549**, 252-265

[14] Martin LW, Crane SP, Chu YH, Holcomb MB, Gajek M, Huijben M, Yang CH, Balke N, Ramesh R. Multiferroics and magnetoelectrics: thin films and nanostructures. J. Phys.: Condens. Matter. 2008; **20**, 434220 (1-13)

[15] Safi R, Shokrollahi H. Physics, chemistry and synthesis methods of nanostructured bismuth ferrite ($BiFeO_3$) as a ferroelectro-magnetic material. Prog. Solid State Chem. 2012; **40**, 6-15

[16] Selbach SM, Einarsrud MA, Tybell T, Grande T. Synthesis of $BiFeO_3$ by Wet Chemical Methods. J. Am. Ceram. Soc. 2007; **90 [11]**, 3430-3434

[17] Xu JH, Ke H, Jia DC, Wang W, Zhou Y. Low-temperature synthesis of $BiFeO_3$ nanopowders via a sol-gel method. J. Alloys Compd. 2009; **472**, 473-477

[18] Shami MY, Awan MS, Anis-ur-Rehman M. Phase pure synthesis of $BiFeO_3$ nanopowders using diverse precursor via co-precipitation method. J. Alloys Compd. 2011; **509**, 10139- 10144

[19] Liu B, Hu B, Du Z, Hydrothermal synthesis and magnetic properties of single-crystalline BiFeO3 nanowires. Chem. Commun., 2011, **47**, 8166–8168

[20] Manzoor A, Afzal AM, Umair M, Ali A, Rizwan M, Yaqoob MZ. Synthesis and characterization of Bismuth ferrite ($BiFeO_3$) nanoparticles by solution evaporation method. J. Magn. Magn. Mater. 2015; **393**, 269-272

[21] Li S, Zhang G, Zheng H, Wang N, Zheng Y, Wang P. Microwave-assisted synthesis of $BiFeO_3$ nanoparticles with high catalytic performance in microwave-enhanced Fenton-like process. RSC Adv., 2016; **6**, 82439-82446

[22] Joshi UA, Jang JS, Borse PH, Lee JS, Microwave synthesis of single-crystalline perovskite $BiFeO_3$ nanocubes for photoelectrode and photocatalytic applications. Appl. Phys. Lett. 2008; **92**, 242106-3

[23] Wu J, Mao S, Ye ZG, Xiea Z, Zheng L. Room-temperature ferromagnetic/ferroelectric $BiFeO_3$ synthesized by a self-catalyzed fast reaction process. J. Mater. Chem. 2010; **20**, 6512-6516

[24] Kim TC, Ojha S, Tian G, Lee SH, Jung HK, Choi JW, Kornblum L, Walker FJ, Ahn CH, Ross CA, Kim DH, Self-assembled multiferroic epitaxial $BiFeO_3$-$CoFe_2O_4$ nanocomposite thin films grown by RF magnetron sputtering. **J. Mater. Chem. C** 2018; **6**, 5552-5561

[25] Zhang L, Cao XF, Ma YL, Chen XT, Xue ZL, Polymer-directed synthesis and magnetic property of nanoparticles-assembled $BiFeO_3$ microrods. J. Solid State Chem. 2010; **183**, 1761-1766

[26] Wei J, Xue D, Xu Y, Photoabsorption characterization and magnetic property of multiferroic $BiFeO_3$ nanotubes synthesized by a facile sol-gel template process. Scripta Mater. 2008; **58**, 45-48

[27] Dutta DP, Jayakumar OD, Tyagi AK, Girija KG, Pillaia CGS, SharmG, Effect of doping on the morphology and multiferroic properties of $BiFeO_3$ nanorods. Nanoscale 2010; **2**, 1149-1154

[28] Gao F, Yuan Y, Wang KF, Chen XY, Chen F, Liu JM, Ren ZF, Preparation and photoabsorption characterization of $BiFeO_3$ nanowires. Appl. Phys. Lett. 2006; **89**, 102506-3

[29] Xie SH, Li JY, Proksch R, Liu YM, Zhou YC, Liu YY, Ou Y, Lan LN, Qiao Y, Nanocrystalline multiferroic $BiFeO_3$ ultrafine fibers by sol-gel based Electrospinning. Appl. Phys. Lett. 2008; **93**, 222904-3

[30] Gupta R, Chaudhary S, Kotnala RK. Interfacial Charge Induced Magnetoelectric Coupling at $BiFeO_3$/$BaTiO_3$ Bilayer Interface. *ACS Appl. Mater. Interf.* 2015; **7(16)**, 8472-8479

[31] Chaturvedi S, Das R, Poddar P, S Kulkarni. Tunable band gap and coercivity of bismuth ferrite-polyaniline core-shell nanoparticles: the role of shell thickness. RSC Adv., 2015, **5**, 23563-23568

[32] Aimon NM, Kim DH, Sun XY, Ross CA. Multiferroic Behavior of Templated $BiFeO_3$-$CoFe_2O_4$ Self-Assembled Nanocomposites. ACS Appl. Mater. Interf. 2015; 7, 2263-2268

[33] Lorenz M, Lazenka V, Schwinkendorf P, Bern F, Ziese M, Modarresi H, Volodin A, Van Bael MJ, Temst K, Vantomme A, Grundmann M. Multiferroic $BaTiO_3$–$BiFeO_3$ composite thin films and multilayers: strain engineering and magnetoelectric coupling. J. Phys. D: Appl. Phys. 2014; **47**, 135303-10

[34] Lorenz M, Wagner G, Lazenka V, Schwinkendorf P, Modarresi H, Van Bael MJ, Vantomme A, Temst K, Oeckler O, Grundmann M. Correlation of magnetoelectric coupling in

multiferroic $BaTiO_3$-$BiFeO_3$ superlattices with oxygen vacancies and antiphase octahedral rotations. Appl. Phys. Lett. 2015; **106**, 012905

[35] Wang J, Neaton JB, Zheng H, Nagarajan V, Ogale SB, Liu B, Viehland D, Vaithyanathan V, Schlom DG, Waghmare UV, Spaldin NA, Rabe KM, Wuttig M, Ramesh R. Epitaxial $BiFeO_3$ Multiferroic Thin Film Heterostructures. Science 2003; **299**, 1719-1722

[36] Kotnala RK, Gupta R, Chaudhary S. Giant magnetoelectric coupling interaction in $BaTiO_3$/$BiFeO_3$/$BaTiO_3$ trilayer multiferroic heterostructures. Appl. Phys. Lett. 2015; **107**, 08290

[37] Kumar N, Shukla A, Kumar N, Choudhary RNP, Kumar A. Structural, electrical, and multiferroic characteristics of lead-free multiferroic: $Bi(Co_{0.5}Ti_{0.5})O_3$-$BiFeO_3$ solid solution. RSC Adv. 2018, **8**, 36939-36950

[38] Chu YH, He Q, Yang CH, Yu P, Martin LW, Shafer P, Ramesh R. Nanoscale Control of Domain Architectures in $BiFeO_3$ Thin Films. Nano Lett. 2009; 9(4), 1727-1730

[39] Park TJ, Papaefthymiou GC, Viescas AJ, Moodenbaugh AR, Wong SS. Size-dependent magnetic properties of single-crystalline multiferroic $BiFeO_3$ nanoparticles. Nano Lett. 2007; **7(3)**, 766-772

[40] Bajpai OP, Mandal S, Ananthakrishnan R, Mandal P, Khastgir D, Chattopadhyay S. Structural Features, Magnetic Properties and Photocatalytic Activity of Bismuth Ferrite Nanoparticles Grafted on Graphene Nanosheets. **New J. Chem.** 2018; **42**, 10712-10723

[41] Amrillah T, Bitla Y, Shin K, Yang T, Hsieh YH, Chiou YY, Liu HJ, Do TH, Su D, Chen YC, Jen SU, Chen LQ, Kim KH, Juang JY, Chu YH. Flexible Multiferroic Bulk Heterojunction with Giant Magnetoelectric Coupling via van der Waals Epitaxy. ACS Nano 2017; **11**, 6122-6130

[42] Catalan G, Scott JF. Physics and Applications of Bismuth Ferrite. Adv. Mater. 2009; **21**, 2463-2485

[43] Yuan X, Shi L, Zhao J, Zhou S, Miao X, Guo J. Tunability of bandgap and magnetism in K and Pb Co-doped $BiFeO_3$ nanoparticles for multiferroic applications: The Role of structural transition and Fe deficiency. *ACS Appl. Nano Mater.* 2019; **2(4)**, 1995-2004

[44] Kumar A, Kumar A, Saha S, Basumatary H, Ranjan R. Ferromagnetism in the multiferroic alloy systems $BiFeO_3$-$BaTiO_3$ and $BiFeO_3$-$SrTiO_3$: Intrinsic or extrinsic?. Appl. Phys. Lett. 2019; **114**, 022902

[45] Jian G, Xue F, Gao M, Guo Y. Orientation dependence of magnetic properties in $BiFeO_3$. J. Magn. Magn. Mater. 2019; **476**, 188-193

[46] Coey JMD, Douvalis AP, Fitzgerald CB, Venkatesan M. Ferromagnetism in Fe-doped SnO_2 thin films. Appl. Phys. Lett. 2004; **84**, 1332

[47] Neel L. Compt. Rend. 1961; **252**, 4075-4080.

[48] Neel L. In Low Temperature Physics. Eds. Gordon and Beach: London, 1962; **411**.

[49] Wu L, Dong C, Chen H, Yao J, Jiang C, Xue D, Varela JA. Hydrothermal synthesis and magnetic properties of bismuth ferrites nanocrystals with various morphology. J. Am. Ceram. Soc. 2012; **95**, 3922-3927

[50] Dutta DP, Mandal BP, Naik R, Lawes G, Tyagi AK. Magnetic, ferroelectric, and magnetocapacitive properties of sonochemically synthesized Sc-doped $BiFeO_3$

Nanoparticles. J. Phys. Chem. C 2013; **117**, 2382-2389

[51] Chen J, Tang Z, Zhao S. Giant magnetoelectric coupling effect in lead-free perovskite $BiFeO_3/Na_{0.5}Bi_{4.5}Ti_4O_{15}$ composite films. J. Alloys Compd. 2017; **712**, 256-262

[52] Pan Q, Chu B. Enhanced magnetoelectric response in bismuth-deficient $BiFeO_3$-$BaTiO_3$ ceramics. J. Appl. Phys. 2019; **125**, 154102

[53] Verma KC, Kotnala RK. Multiferroic magnetoelectric coupling and relaxor ferroelectric behavior in $0.7BiFeO_3$-$0.3BaTiO_3$ nanocrystals. Solid State Commun. 2011; **151**, 920-923

[54] Shirolkar MM, Li J, Dong X, Li M, Wang H. Controlling the ferroelectric and resistive switching properties of $BiFeO_3$ thin film prepared using sub - 5 nm dimension nanoparticles. **Phys. Chem. Chem. Phys.** 2017; **19**, 26085-26097

[55] Verma KC, Singh D, Kumar S, Kotnala RK. Multiferroic effects in $MFe_2O_4/BaTiO_3$ (M = Mn, Co, Ni, Zn) Nanocomposites. J. Alloys Compd. 2017; **709**, 344-355

[56] Verma KC, Singh M, Kotnala RK, Goyal N. Magnetic field control of polarization/capacitance/voltage/resistance through lattice strain in $BaTiO_3$-$CoFe_2O_4$ multiferroic nanocomposite. J. Magn. Magn. Mater. 2019; **469**, 483-493

[57] Pan L, Yuan Q, Liao Z, Qin L, Bi J, Gao D, Wu J, Wu H, Ye ZG. Superior room-temperature magnetic field-dependent magnetoelectric effect in $BiFeO_3$-based multiferroic. J. Alloys Compd. 2018; **762**, 184-189

[58] Jochum JK, Lorenz M, Gunnlaugsson HP, Patzig C, Hoche T, Grundmann M, Vantomme A, Temst K, Van Bael MJ, Lazenka V. Impact of magnetization and hyperfine field distribution on high magnetoelectric coupling strength in $BaTiO_3$-$BiFeO_3$ multilayers. **Nanoscale**, 2018, **10**, 5574-5580

[59] Lazenka V, Lorez M, Modarresi H, Bisht M, Ruffer R, Bonholzer M, Grundmann M, Van Bael MJ, Vantomme A, Temst K. Magnetic spin structure and magnetoelectric coupling in $BiFeO_3$-$BaTiO_3$ multilayer. Appl. Phys. Lett. 2015; **106**, 082904

[60] Verma KC, Kotnala RK. Multiferroic approach for Cr,Mn,Fe,Co, Ni,Cu substituted $BaTiO_3$ nanoparticles. Mater. Res. Exp.2016; **3**, 055006

[61] Yang B, Jin L, Wei R, Tang X, Hu L, Tong P, Yang J, Song W, Dai J, Zhu X, Sun Y, Zhang S, Wang X, Cheng Z, Chemical Solution Route for High-Quality Multiferroic $BiFeO_3$ Thin Films. Small 2019; **doi.org/10.1002/smll.201903663**, 1903663-19

[62] Sun B, Han P, Zhao W, Liu Y, Chen P, White-Light-Controlled Magnetic and Ferroelectric Properties in Multiferroic $BiFeO_3$ Square Nanosheets. J. Phys. Chem. C 2014; **118**, 18814-18819

[63] Kolte J, Daryapurkar AS, Agarwal M, Gulwade DD, Gopalan P, Magnetoelectric properties of microwave sintered $BiFeO_3$ and $Bi_{0.90}La_{0.10}Fe_{0.95}Mn_{0.05}O_3$ Nanoceramics. Mater. Chem. Phys. 2017; **193**, 253-259

[64] Rout J, Choudhary RNP, Structural transformation and multiferroic properties of Ba-Mn co-doped $BiFeO_3$. Phys. Lett. A 2016; **380**, 288-292

[65] Banerjee M, Mukherjee A, Banerjee A, Das D, Basu S, Enhancement of multiferroic properties and unusual magnetic phase transition in Eu doped bismuth ferrite nanoparticles. New J. Chem. 2017; **41**, 10985-10991

[66] Mukherjee A, Basu S, Manna PK, Yusuf SM, Pal M, Giant magnetodielectric and enhanced multiferroic properties of Sm doped bismuth ferrite nanoparticles. J. Mater. Chem. C 2014; **2**, 5885-5891

[67] Dutta DP, Mandal BP, Mukadam MD, Yusuf SM, Tyagi AK, Improved magnetic and ferroelectric properties of Sc and Ti codoped multiferroic nano $BiFeO_3$ prepared via sonochemical synthesis. Dalton Trans. 2014; **43**, 7838-7846

[68] Puli VS, Pradhan DK, Gollapudi S, Coondoo I, Panwar N, Adireddy S, Chrisey DB, Katiyar RS, Magnetoelectric coupling effect in transition metal modified polycrystalline $BiFeO_3$ thin films. J. Mag. Magn. Mater. 2014; **369**, 9-13

[69] Yang S, Ma G, Xu L, Deng C, Wang X, Improved ferroelectric properties and band-gap tuning in $BiFeO_3$ films via substitution of Mn. RSC Adv. 2019; **9**, 29238-29245

[70] Jena AK, Satapathy S, Mohanty J, Magnetic properties and oxygen migration induced resistive switching effect in Y substituted multiferroic bismuth ferrite. Phys. Chem. Chem. Phys. 2019; **21**, 15854-15860

[71] Marzouki A, Loyau V, Gemeiner P, Bessais L, Dkhil B, Megrich A, Increase of magnetic and magnetoelectric properties in Co/Mn co-doped BiFeO3 multiferroic. J. Mag. Magn. Mater. 2020; **498**, 166137

[72] Li CX, Yang B, Zhang ST, Zhang R, Sun Y, Zhang HJ, Cao WW, Enhanced Multiferroic and Magnetocapacitive Properties of (1–x) $Ba_{0.7}Ca_{0.3}TiO_3$–x$BiFeO_3$ Ceramics. J. Am. Ceram. Soc. 2014; **97 [3]**, 816-825

[73] Behera C, Choudhary RNP, Das PR, Structural, Electrical and Multiferroic Characteristics of Thermo-mechanically Fabricated $BiFeO_3$-(BaSr)TiO_3 Solid Solutions. *Mater. Res. Expr.* 2018; **5**, 056301

[74] Qi X, Zhang M, Zhang X, Gu Y, Zhu H, Yang W, Li Y, Compositional dependence of ferromagnetic and magnetoelectric effect properties in $BaTiO_3$–$BiFeO_3$–$LaFeO_3$ solid solutions. RSC Adv. 2017; 7, 51801-51806

[75] Ummer RP, Thevenot RC, Rouxel D, Thomas S, Kalarikkal N, Electric, magnetic, piezoelectric and magnetoelectric studies of phase pure ($BiFeO_3$-$NaNbO_3$)-(P(VDF-TrFE)) nanocomposite films prepared by spin coating. RSC Adv. 2016; **6**, 28069-28080

[76] Calisir I, Amirov A A, Kleppe AK, Hall DA, Optimisation of functional properties in lead-free $BiFeO_3$-$BaTiO_3$ ceramics through La^{3+} substitution strategy. **J. Mater. Chem. A** 2018; **6**, 5378-5397

[77] Gupta R, Shah J, Chaudhary S, Singh S, Kotnala RK, Magnetoelectric coupling-induced anisotropy in multiferroic nanocomposite (1 - x) $BiFeO_3$–x$BaTiO_3$. J Nanopart Res 2013; **15**, 2004

[78] Remya KP, Rajalakshmi R, Ponpandian N, Development of $BiFeO_3$/$MnFe_2O_4$ ferrite nanocomposites with enhanced magnetic and electrical properties. Nanoscale Adv. 2020; **2**, 2968-2976

[79] Jin C, Geng W, Wang L, Han W, Zheng D, Hu S, Ye M, Xu Z, Ji Y, Zhao J, Chen Z, Wang G, Tang Y, Zhu Y, Mac X, Chen L, Tuning Ferroelectricity and Ferromagnetism in $BiFeO_3$/$BiMnO_3$ Superlattices. Nanoscale 2020; **12**, 9810-9816

[80] Sheoran N, Kumar A, Kumar V, Chahar M, Banerjee A, Structural and multiferroic properties of $BiFeO_3$/$MgLa_{0.025}Fe_{1.975}O_4$ nanocomposite synthesized by sol-gel auto combustion route. J. Mater. Sci. 2020; **31**, 2788

[81] Sufyan M, Atiq S, Abbas SK, Younis M, Riaz S, Naseem S, Magnetically driven robust polarization in (1-x) $BiFeO_3$-$x$$PbTiO_3$ multiferroic composites. 2019; **238**, 10-12

[82] Jain A, Wang YG, Wang N, Li Y, Wang FL, Existence of heterogeneous phases with significant improvement in electrical and magnetoelectric properties of $BaFe_{12}O_{19}$/$BiFeO_3$ multiferroic ceramic composites. Cer. Int. 2019; **45**, 22889-22898

[83] Alam M, Talukdar S, Mandal K, Multiferroic properties of bilayered $BiFeO_3$/$CoFe_2O_4$ nano-hollowspheres. Mater. Lett. 2018; **210**, 80-83

[84] Shariq M, Kaur D, Chandel VS, Jain PK, Florence S, Sharma M, Hussain S, Study of Structural, Magnetic and Optical Properties of $BiFeO_3$-$PbTiO_3$ Multiferroic Composites. Arab. J. Sci. Engg. 2019; **44**, 613-621

Chapter 8

Investigation of Structural, Microstructural, Dielectrical and Magnetic Properties of Bi3+ Doped Manganese Spinel Ferrite Nanoparticles for Photonic Applications

V. Jagadeesha Angadi, H.R. Lakshmiprasanna and K. Manjunatha

Abstract

The structural, microstructural, and magnetic properties of $Mn_{1-x}Bi_xFe_2O_4$ (where x = 0.0, 0.05, 0.1, 0.15, and 0.2) nanoparticles prepared by solution combustion method were investigated. Rietveld-refined X-ray diffraction patterns confirm the single-phase formation with space group Fd3m having spinel cubic structure. The porous nature of the samples was confirmed by scanning electron microscopy (SEM). Composition values of the theoretical stoichiometry and energy-dispersive spectroscopy (EDS) composition values are well matched for all samples. The dielectric parameters such as real part of dielectric constant, imaginary part of dielectric constant, and dielectric loss tangent decrease with the increase in frequency. The AC conductivity increases with increase in the Bi^{3+} concentration. The real part of complex impedance decreases with the increase in frequency. Cole-Cole plots reveal that one semicircle was obtained for each of the samples. The real and imaginary parts of electric modulus vary with frequency. The magnetic hysteresis curves of all samples reveal the soft magnetic material nature. We observed S esteems began uniquely from the higher superparamagnetic, we would have watched the monotonic decrease in S with increase in Bi^{3+} concentration. Furthermore, the magnetic parameters were estimated.

Keywords: Maxwell-Wagner interfacial type of polarization, Cole-Cole plots, soft magnetic material, solution combustion method

1. Introduction

Nowadays, manganese (Mn) ferrite nanoparticles have been used for great potential applications such as absorbing electromagnetic waves, storage media, and water treatment catalysts [1–4]. Many researchers focused on the fabrication of spinel ferrite nanoparticles due to their exceptional electrical and magnetic properties [5–9]. To a progressively essential degree, the expanded scaling down and data storage devices of new devices require the utilization of nanosized magnetic attractive particles [10, 11]. Manganese spinel ferrite nanoparticles have gotten expanding consideration for their surprising magnetic properties, such as moderate saturation magnetization (M_s) and low coercivity with a mechanical hardness, high permeability, and good chemical stability [12, 13]. The significance of thick manganese (Mn) spinel ferrites additionally lies in their potential application, for example, transformers, core materials for coils, information and communication devices, and so forth. The $MnFe_2O_4$ nanoparticles have a spinel cubic structure having the general formula AB_2O_4, where the A sites are occupied by the Mn^{2+} ions and B sites by the Fe^{3+} ions. In a unit cell of manganese ferrites, there are 8 divalent Mn^{2+} ions, 16 trivalent Fe^{3+} ions, and 32 oxygen ions [14].

The spinel $MnFe_2O_4$ nanoparticles were prepared by synthesis methods such as sonochemical technique, co-precipitation, sol-gel method, and solution combustion method [15–17]. In the current work, the solution combustion method is often chosen because it is safe, cheap, and easy and it requires less time and energy for the preparation of nanoparticles. It also produces a high-purity product. In this method, the crystallite size, particle size, purity, surface area, degree of nature, and porosity can be influenced by the fuel (F) type [18]. However, for the preparation of nanoparticles, there is no requirement for complex tools, high-temperature furnaces, and reaction chambers in the synthesis process.

In the present work, the Bi^{3+}-substituted $MnFe_2O_4$ spinel nanoparticles were prepared by a solution combustion method using a mixture of fuels such as glucose and urea. The structural, microstructural, dielectrical, and magnetic properties of Bi-doped manganese spinel ferrite nanoparticles were synthesized and investigated through XRD, SEM, EDS, and dielectric and vibrating sample magnetometer (VSM) for photonic applications.

2. Experimental details

2.1 Materials

For the preparation of $Mn_{1-x}Bi_xFe_2O_4$ (x = 0.0, 0.05, 0.1, 0.15, and 0.2) nanoparticles, $Mn(NO_3)_2.4H_2O$ (manganese nitrate) (Sigma-Aldrich; molecular weight 251.01 g/mol, purity 99.0%), $Bi(NO_3)_3{\cdot}5H_2O$ (bismuth nitrate) (Leica, Sigma-Aldrich; purity 98 + %, molecular weight 485.07 g/mol), and $Fe(NO_3)_2.9H_2O$ (ferric nitrate) (Merck; purity 99.0%, molecular weight 404 g/mol) are used as oxidizers (O) and NH_2CONH_2 (carbamide) (Fisher Scientific; purity 99.0%, molecular weight 60.06 g/mol) and $C_6H_{12}O_6$ (glucose) (Glucomin-D; purity 99.4%, molecular weight 180.156 g/mol) as fuels. Further based on the oxidizer's and fuel's valences of the compounds, the oxidizer-and-fuel ratio is maintained to unity.

2.2 Synthesis method

The $Mn_{1-x}Bi_xFe_2O_4$ (x = 0.0, 0.05, 0.1, 0.15, and 0.2) spinel nanoparticles were prepared by the solution combustion method using a stoichiometric amount of O and a mixture of F proportions taken as one. Stoichiometric amounts of manganese nitrate ($Mn(NO_3)_2{\cdot}4H_2O$), bismuth nitrate ($Bi(NO_3)_3{\cdot}5H_2O$), ferric nitrate ($Fe(NO_3)_3{\cdot}9H_2O$), carbamide (NH_2CONH_2), and glucose ($C_6H_{12}O_6$) are carried and put in a 500-ml Borosil glass beaker, which is then diluted with a double distilled water. The mixture solution was kept on magnetic stirrer about 60 min to get a homogeneous solution. This homogeneous solution was kept in a pre-warmed suppress heater at 450°C. Within 20 min, the burning process was finished. Then, the obtained powder was grained about 30 min using agate mortar and pestle. The flow chart of the $Mn_{1-x}Bi_xFe_2O_4$ (where x = 0.0, 0.05, 0.1, 0.15, and 0.2) solution combustion is shown in **Figure 1**.

2.3 Characterizations

The XRD patterns were gotten by using CuK_α radiation having a wavelength (λ) = 1.5406 Å, and the 2θ values run from 10° to 80° under the step size of 0.02°. The crystalline nature and phase structure can be understood from the XRD patterns. The SEM and EDS micrographs were collected by using JEOL (model JSM-840) to obtain the surface morphology and elemental composition of the samples. To study the electrical properties, the samples are made into pellets using a hydraulic press under the 5-ton pressure using an 11-mm die set. The prepared pellets are sintered at 650°C for 3 h. The sintered pellets are silver coated for a good electrical contact. The electrical properties of all samples are tested at room temperature in the frequency up to 20 MHz using the Wayne Kerr 6500B Series Impedance Analyzer. The magnetic properties were measured at room temperature using a physical properties measurement system (PPMS; Quantum Design, Inc.) up to a field of 20 kOe.

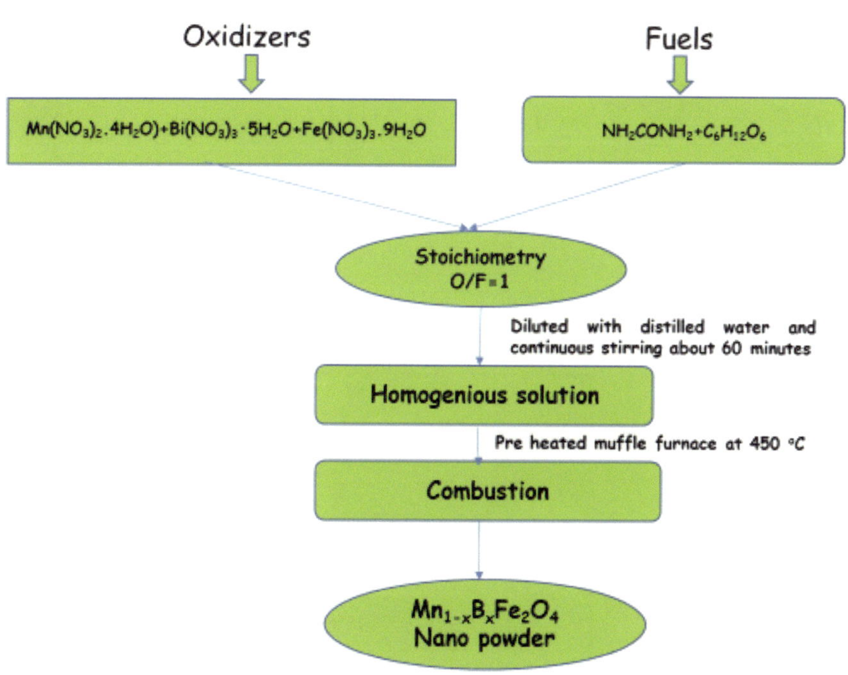

Figure 1.
The flow chart of the solution combustion method for $Mn_{1-x}Bi_xFe_2O_4$ (where x = 0.0, 0.05, 0.1, 0.15, and 0.2) nanoparticles.

3. Results and discussions

3.1 Phase, crystallinity, and structural elucidation

Figure 2 depicts the refined XRD patterns of $Mn_{1-x}Bi_xFe_2O_4$ (where x = 0.0, 0.05, 0.1, 0.15, and 0.2) as synthesized ferrite samples were plotted using a WinPLOTR program. The Bragg diffraction peaks corresponding to angular positions were indexed as (220) (311) (222) (400) (331) (422) (511) (440). Rietveld refinement was employed using a FullProf program for structural refinements. The parameters obtained from refinement, such as weighted (R_p) and unweighted (R_{wp})

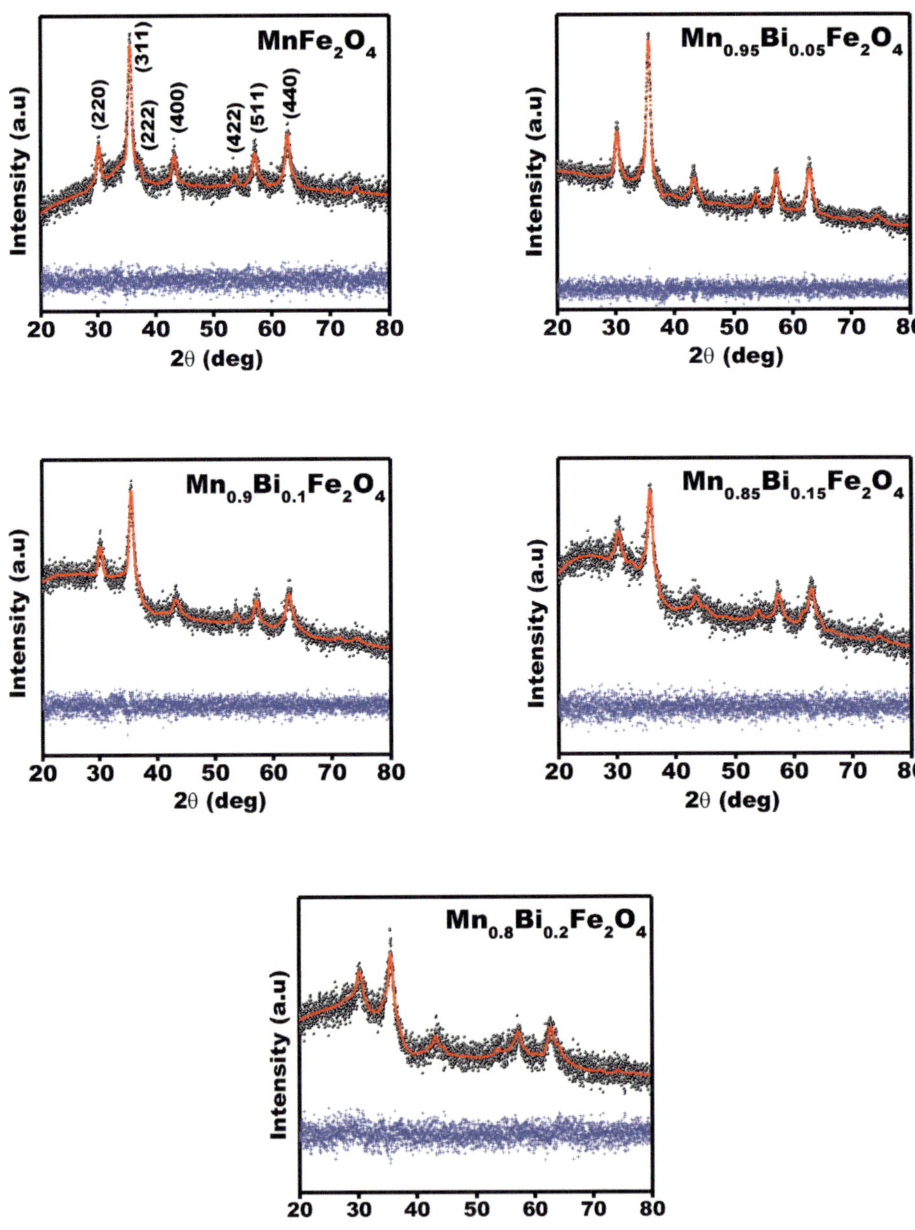

Figure 2.
Rietveld-refined XRD patterns of $Mn_{1-x}Bi_xFe_2O_4$ (where x = 0.0, 0.05, 0.1, 0.15, and 0.2) nanoparticles.

profile factor ratio (R_{wp}/R_p), which is less than one, and goodness-of-fit factor (χ^2), which is low, are given in **Table 1**. The refined XRD patterns confirm the single phase with a spinel cubic structure having a space further without any additional impurity peaks. The indexed (hkl) values are well matched with the Joint Committee of Powder Diffraction Standards (JCPDS) card number 74-2400. **Figure 3** clearly shows that the lattice parameters increase with the increase of Bi^{3+} concentration due to the difference in ionic radius of Mn^{2+} and Bi^{3+} ions. The ionic radius of Mn^{2+} (0.83 Å) is less than the ionic radius of Bi^{3+} (1.03 Å). The average crystallite size was calculated by using the equation below [19]:

$$D = \frac{k\lambda}{\beta cos\theta}\, nm \tag{1}$$

where k is the Scherrer constant (k = 0.9), λ is the wavelength of Cu-k_α radiation, β is the full width half maximum, and θ is the angle of diffraction. The estimated crystallite size decreases with the increase of Bi^{3+} concentration (**Figure 3**). Further internal strain, X-ray density, and hopping length were increased with Bi^{3+} concentration increasing (**Figure 4**). The lattice parameters, average crystallite size, internal strain, X-ray density, and hopping lengths were tabulated in **Table 1**.

Bi^{3+} content	Lattice parameters (Å)	Crystallite size D in (nm)	Volume (Å^3)	Internal strain ∈ 10^{-2} (%)	X-ray density (Δ) g/cm^3	Hoping length (Å)	
						L_A	L_B
0.0	8.3612	13.7	584.52	88	5.2406	3.6025	2.9561
0.05	8.3658	11.9	585.49	95	5.4068	3.6225	2.9577
0.10	8.3845	10.8	589.42	96	5.5443	3.6306	2.9643
0.15	8.3874	10.6	590.04	105	5.7117	3.6319	2.9654
0.20	8.3932	9.2	591.26	131	5.8731	3.6343	2.9674

Table 1.
Crystallite size, lattice constant, strain, X-ray density, magnetic hoping length data of $Mn_{1-x}Bi_xFe_2O_4$ (where x = 0.0, 0.05, 0.1, 0.15, and 0.2) nanoparticles.

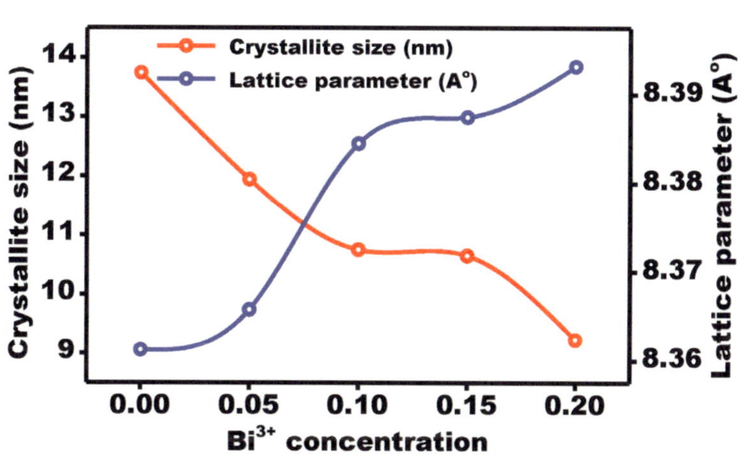

Figure 3.
Variation of crystallite size (nm) and lattice parameter (a°) with Bi^{3+} doping concentration of $Mn_{1-x}Bi_xFe_2O_4$ (where x = 0.0, 0.05, 0.1, 0.15, and 0.2) nanoparticles.

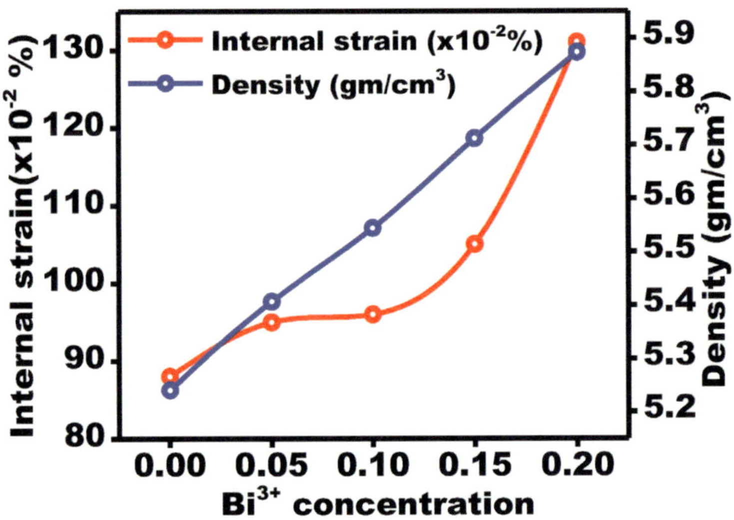

Figure 4.
Variation of internal strain (%) and density (gm/cm^3) of (where x = 0.0, 0.05, 0.1, 0.15, and 0.2) nanoparticles with Bi^{3+} doping concentration.

3.2 SEM and EDS

Figures 5 and **6** show the SEM and EDS of Bi^{3+}-doped $MnFe_2O_4$ (where x = 0.0, 0.05, 0.1, 0.15, and 0.2) ferrite samples synthesized, respectively. All SEM micrographs reveal the highly porous nature of the samples [20]. All the samples show the dry foamy powder due to the combustion. The average grain size cannot be observed due to the high porous nature of the sample. The density of samples increases with Bi^{3+} concentration. The EDS images clearly reveal that Mn, Fe, and O peaks are clearly visible in all samples. The Bi^{3+} peak is visible in all samples except at 0 mol % Bi^{3+} sample. The stoichiometry of constituent elements present in the $Mn_{1-x}Bi_xFe_2O_4$ with x = 0.0, 0.05, 0.1, 0.15, and 0.2 ferrite nanoparticles estimated from EDS spectrum is shown in **Table 2**. The estimated stoichiometry as listed in **Table 3** is very near to the composition of theoretical values.

3.3 Dielectric studies

3.3.1 Real (ε') and imaginary (ε'') parts of dielectric constant

The real (ε′) and imaginary (ε′′) parts of the dielectric constant as a function of frequency at room temperature for $Mn_{1-x}Bi_xFe_2O_4$ (where x = 0.0, 0.05, 0.1, 0.15, and 0.2) nanoparticles are depicted in **Figure 7(a)** and **Figure 7(b)**, respectively. At low frequency, the real and imaginary parts of dielectric constants show the dispersion for all samples. The dielectric constant values sharply decrease with increasing in frequency, and it is due to Maxwell-Wagner interfacial-type polarization according to Koop's phenomenological theory [21]. At the lower-frequency region, the interfacial polarization contributes to the dispersion. A systematic variation is observed due to the dispersion of dielectric constant with substitution. At the higher-frequency region, it cannot follow the alternating electric field, so real and imaginary parts of dielectric constants are independent of frequency at that region. In ferrites, the maximum dielectric constant is observed in samples having higher Fe^{2+} content. At the lower-frequency region, the hopping (exchange

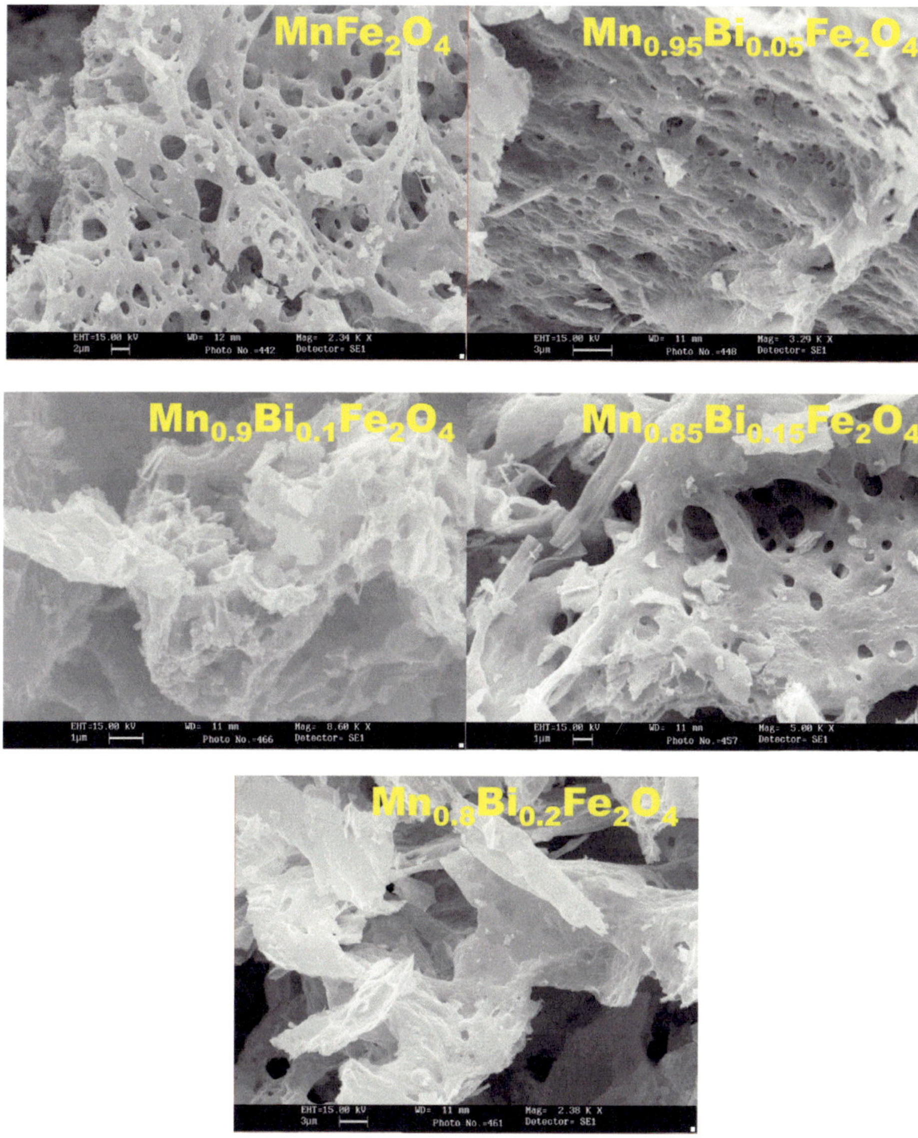

Figure 5.
SEM micrograph images of $Mn_{1\text{-}x}Bi_xFe_2O_4$ ferrite nanoparticles with x = 0.0 ($MnFe_2O_4$), x = 0.05 ($Mn_{0.95}Bi_{0.05}Fe_2O_4$), x = 0.1 ($Mn_{0.9}Bi_{0.1}Fe_2O_4$), x = 0.15 ($Mn_{0.85}Bi_{0.15}Fe_2O_4$), and x = 0.2 ($Mn_{0.8}Bi_{0.2}Fe_2O_4$).

of electron) between Fe^{2+} and Fe^{3+} at octahedral sites follows the alternating electric field. Many literatures on AB_2O_4 ferrites indicate that high-temperature sintering leads to the exchange of the Fe^{3+} ions to Fe^{2+} ions at octahedral sites to some degree [22]. However, the hopping process, involving between the Fe^{3+} and Fe^{2+} ions, is attributed to the dielectric properties of the samples. Thus, the increase of Bi^{3+} concentration results in the increase of dielectric constant. The grain size also influences the dielectric properties, since, in ferrite samples, the dielectric constant of the poorly conducting grain boundary is less than the dielectric constant of highly conducting grains. Thus, size reduction introduces more grain boundaries and hence increases changes in electron polarization at the grain boundaries. Since the size decreases with bismuth substitution increasing, it also increases the dielectric constant.

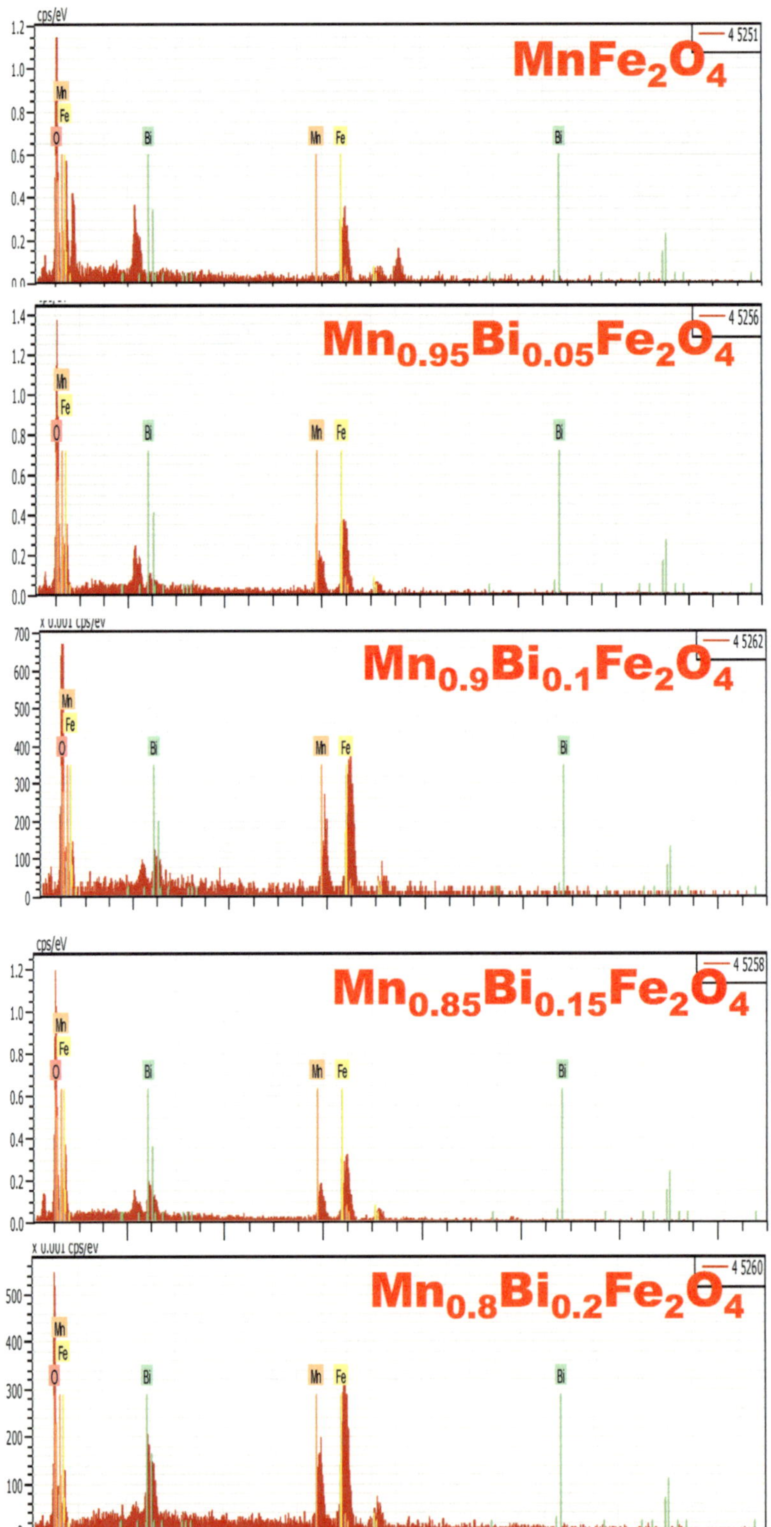

Figure 6.
EDS patterns of $Mn_{1-x}Bi_xFe_2O_4$ ferrite nanoparticles with $x = 0.0$ ($MnFe_2O_4$), $x = 0.05$ ($Mn_{0.95}Bi_{0.05}Fe_2O_4$), $x = 0.1$ ($Mn_{0.9}Bi_{0.1}Fe_2O_4$), 0.15 ($Mn_{0.85}Bi_{0.15}Fe_2O_4$), and $x = 0.2$ ($Mn_{0.8}Bi_{0.2}Fe_2O_4$).

x	Atomic abundance of elements (%)				
	Mn	Bi	Fe	O	Total
0.0	17.92	0.0	32.13	49.96	100
0.05	20.45	0.96	29.07	49.52	100
0.1	18.59	1.75	30.53	49.12	100
0.15	19.37	3.11	29.08	48.44	100
0.2	20.02	4.30	27.83	47.85	100

Table 2.
The stoichiometry of constituent elements present in the $Mn_{1-x}Bi_xFe_2O_4$ (where x = 0.0, 0.05, 0.1, 0.15, and 0.2) nanoparticles estimated from EDS spectrum.

Bi^{3+} concentration (x)	Element	Composition of theoretical values	Composition from EDX analysis
	Mn	1	0.99
0	Bi	0	0
	Fe	1	1
	Mn	0.95	0.955
0.05	Bi	0.05	0.045
	Fe	1	1
	Mn	0.9	0.914
0.1	Bi	0.1	0.086
	Fe	1	1
	Mn	0.85	0.862
0.15	Bi	0.15	0.138
	Fe	1	1
	Mn	0.8	0.823
0.2	Bi	0.2	0.177
	Fe	1	1

Table 3.
Summary of EDS analysis of $Mn_{1-x}Bi_xFe_2O_4$ (where x = 0.0, 0.05, 0.1, 0.15, and 0.2) nanoparticles.

3.3.2 Dielectric loss tangent (tan δ)

The dielectric loss tangent as a function of frequency at room temperature for $Mn_{1-x}Bi_xFe_2O_4$ (where x = 0.0, 0.05, 0.1, 0.15, and 0.2) nanoparticles is shown in **Figure 8**. The figure shows the dielectric loss tangent decreases with the increase of the frequency. The dielectric loss tangent is a result of the string in polarization in accordance with the applied external electric field. The materials with high dielectric loss tangent are engaged for manufacturing high-frequency heating devices. The tan δ was calculated using the relation $\tan \delta = \varepsilon'/\varepsilon''$, where ε' and ε'' are the real and imaginary parts of the dielectric constant, respectively [23].

The tan δ decreases with the increase of the frequency, and then it becomes constant at higher frequencies for all samples. The dielectric loss tangent depends on a number of factors, such as homogeneity, stoichiometry, synthesis method,

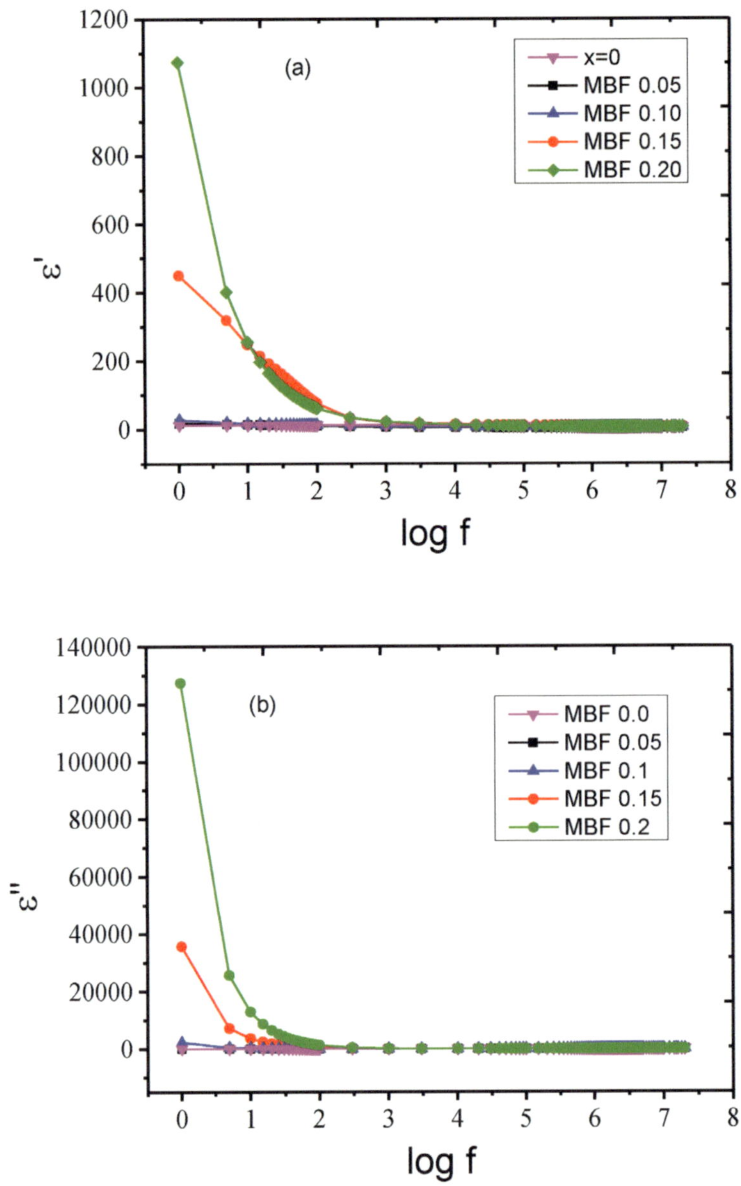

Figure 7.
(a) The real (ε') and (b) imaginary (ε'') parts of the dielectric constant as a function of frequency at room temperature, respectively, for $Mn_{1-x}Bi_xFe_2O_4$ (where x = 0.0, 0.05, 0.1, 0.15, and 0.2) nanoparticles.

ferric content, and composition. The frequency increases with the decrease of the dielectric loss tangent; it may be characterized by the Maxwell-Wagner polarization and conduction mechanism in ferrites.

3.3.3 AC conductivity (σ_{ac})

The variation of σ_{ac} as a function of frequency at room temperature for $Mn_{1-x}Bi_xFe_2O_4$ (where x = 0.0, 0.05, 0.1, 0.15, and 0.2) nanoparticles is shown in **Figure 9**. From the figure, it may well be seen that at first the conductivity stays steady in the low frequencies, bit by bit increments in the middle of the road frequencies, and afterward shows scattering for high frequencies in consistence with the Jonscher power law. The AC conductivity was calculated by using the equation

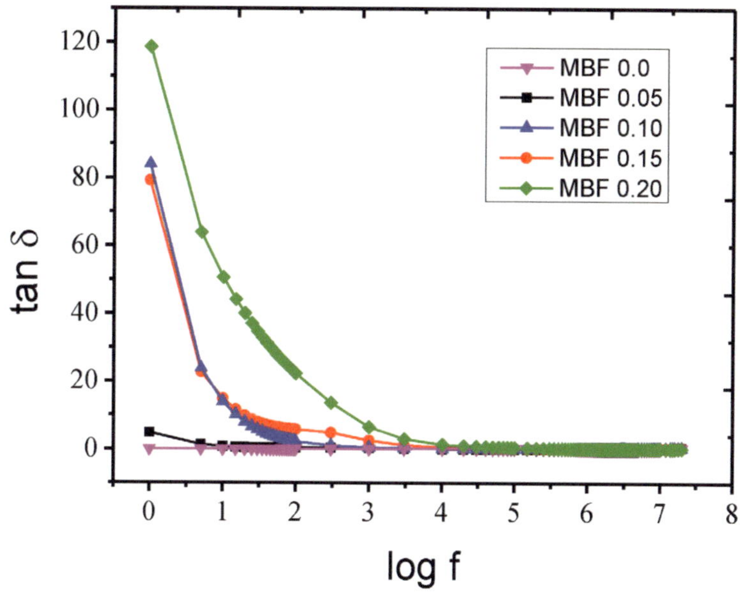

Figure 8.
The dielectric loss tangent as a function of frequency at room temperature for $Mn_{1-x}Bi_xFe_2O_4$ (where x = 0.0, 0.05, 0.1, 0.15, and 0.2) nanoparticles.

$\sigma_{ac} = 2\pi f \varepsilon_0 \times \tan\delta$, where ε_0 is the permittivity of free space, f is the frequency, and tan δ is the dielectric loss tangent. Here, the figure shows that σ_{ac} increases with the increase of the frequency according to the hopping model [24]. In the lower frequencies, grain limits are increasingly adequate; thus the recurrence-free conducts are accomplished. Be that as it may, at high-frequency system, the increase in σ_{ac} is credited to the increased hopping of charge bearers between the Fe^{2+} and Fe^{3+} particles and Mn^{2+} and Mn^{3+} particles at the octahedral site and further as a

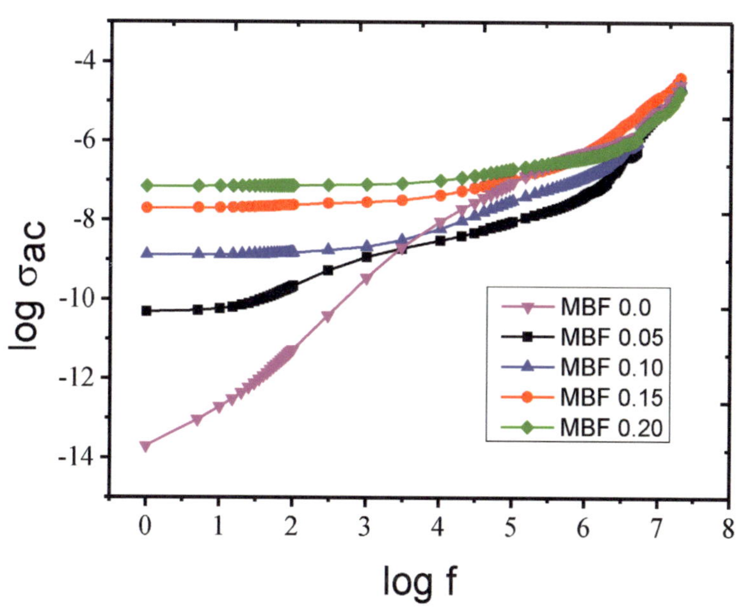

Figure 9.
The AC conductivity as a function of frequency at room temperature for $Mn_{1-x}Bi_xFe_2O_4$ (where x = 0.0, 0.05, 0.1, 0.15, and 0.2) nanoparticles.

result of grain impact. With increasing substitution of Bi^{3+} by Mn^{2+} particles, it very well may be seen that at first conductivity diminishes because of the decrease in the quantity of Mn^{2+}-Mn^{3+} particles which confines the portability of charge transporters at the octahedral destinations.

3.3.4 Real and imaginary parts of complex impedance

Figure 10(a) and **(b)** shows the real and imaginary parts of complex impedance as a function of frequency at room temperature for $Mn_{1-x}Bi_xFe_2O_4$ (where x = 0.0, 0.05, 0.1, 0.15, and 0.2) nanoparticles, respectively. These types of frequency dependence complex impedance have been used for the study of the contribution of grains and grain boundaries on the electrical properties of nanoferrites. The obtained results showed the complex impedance versus frequency for the cobalt

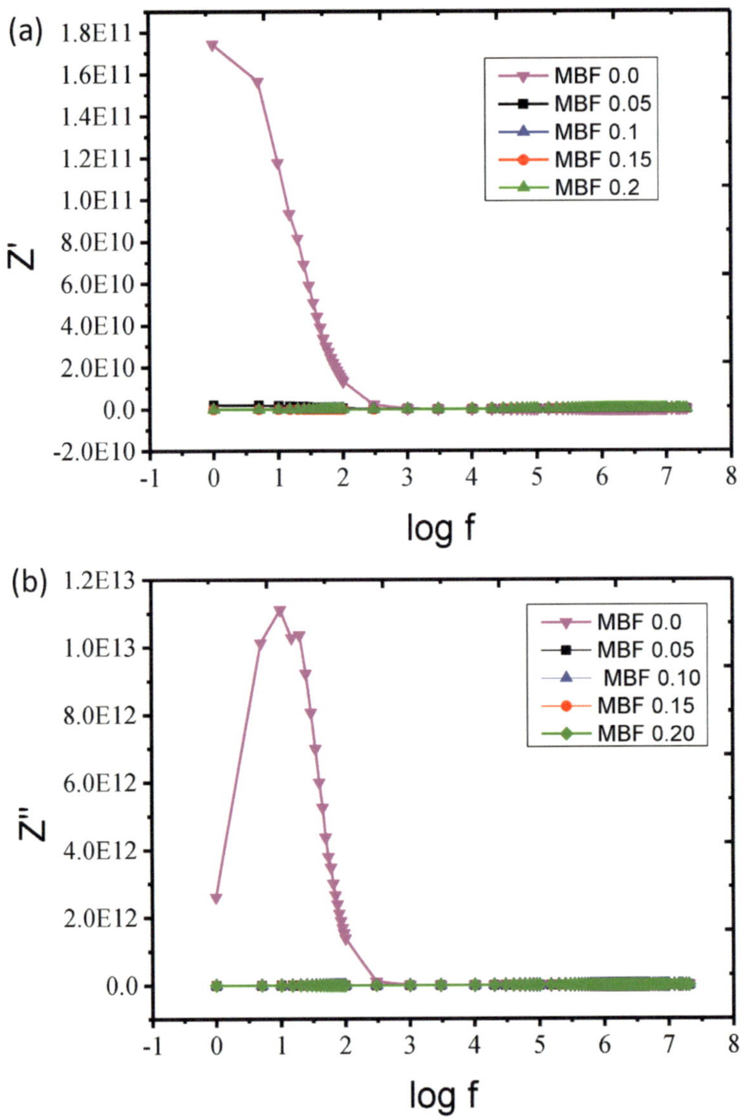

Figure 10.
(a) and (b) The real and imaginary parts of complex impedance as a function of frequency at room temperature for $Mn_{1-x}Bi_xFe_2O_4$ (where x = 0.0, 0.05, 0.1, 0.15, and 0.2) nanoparticles.

ferrites. **Figure 10(a)** indicates that the magnitude of the real part of dielectric constant Z′ decreased with the increase of frequency. The real part of dielectric constant curves tended to merge in the high-frequency region. Such behavior is related to the effect of space charge polarization in the material [25]. **Figure 10(b)** indicates that the imaginary part of dielectric constant Z′′ reached a maximum value at the low-frequency region and then decreased with further increasing of the frequency, approaching a small value at high-frequency region. Further, it was observed that the height of the peak curve decreased with the concentration increasing, indicating a decrease in the relaxation time. The observed relaxation peak in the magnitude of Z′′ demonstrated the existence of space charge relaxation.

3.3.5 Cole-Cole plot

Figure 11 shows the Cole-Cole plots of $Mn_{1-x}Bi_xFe_2O_4$ (where x = 0.0, 0.05, 0.1, 0.15, and 0.2) nanoparticles at room temperature. Cole-Cole plot shows the real part complex impedance (Z′) and the imaginary part of complex impedance (Z′′) along x-axis and y-axis, respectively. **Figure 11** clearly shows one semicircle was obtained for each of the samples, due to the grain boundary effect. The effects of grain boundary of the material are contributed by the parallel combination of grain boundary resistance and capacitance. However, the radius of curvature of each semicircular arc is reasonably shifted from the magnitude of Z′ when the concentration increased, which is attributed to the presence of a non-Debye type of relaxation phenomenon [26]. Further, the radius of curvature of each semicircular arc is indirectly proportional to the electrical conductance of the material and directly proportional to the electrical resistance. The position of the semicircular arc depends on its relaxation times.

3.3.6 The real (M′) and imaginary (M″) parts of electric modulus

Figure 12(a) and **(b)** shows the real (M′) and imaginary (M′′) parts of electric modulus as a function of frequency at room temperature for $Mn_{1-x}Bi_xFe_2O_4$ (where

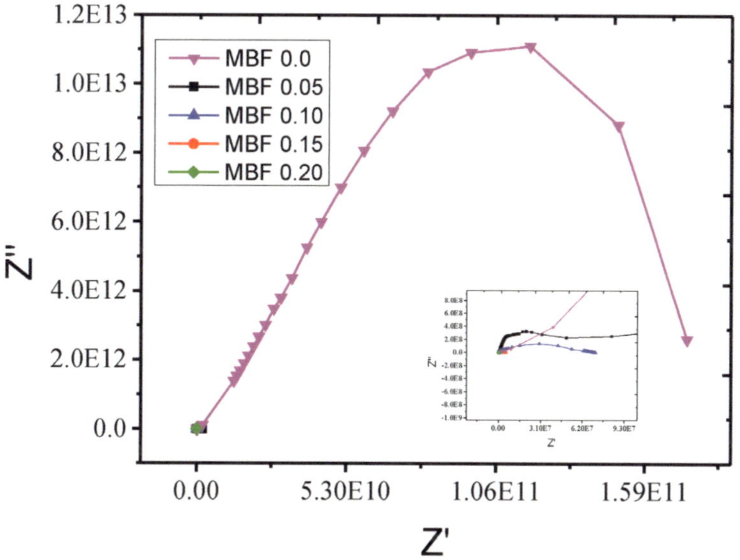

Figure 11.
The Cole-Cole plots of $Mn_{1-x}Bi_xFe_2O_4$ (where x = 0.0, 0.05, 0.1, 0.15, and 0.2) nanoparticles at room temperature.

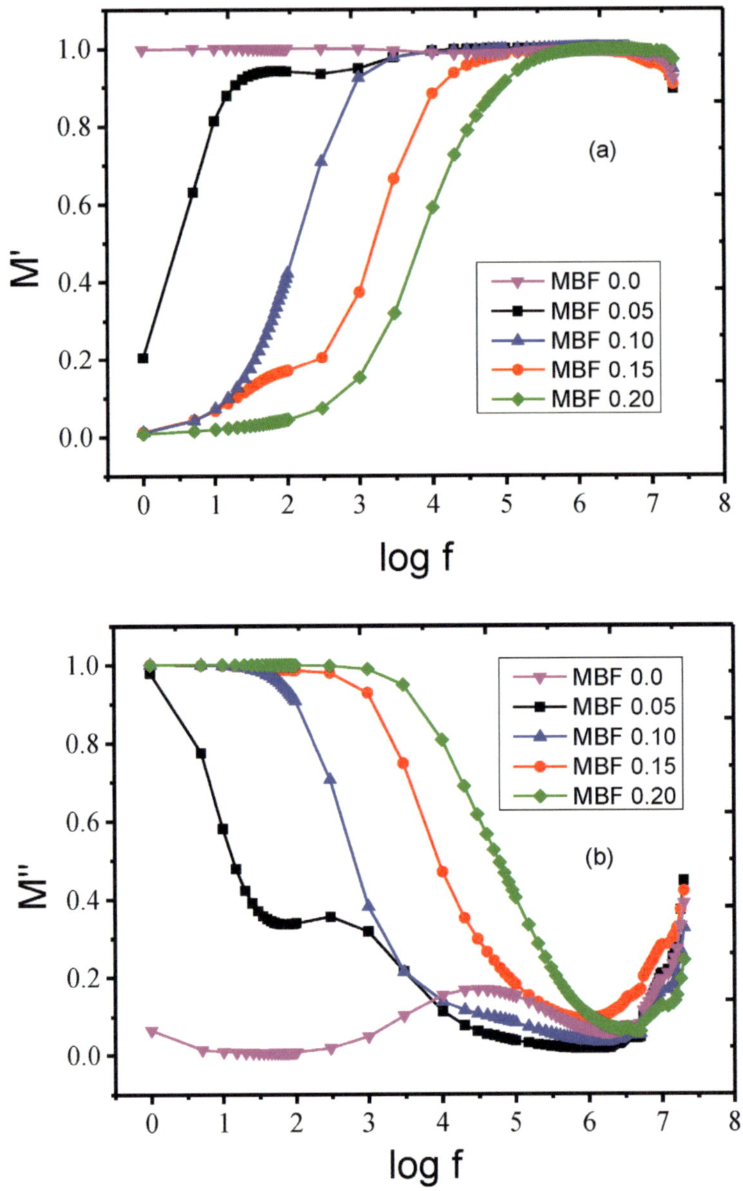

Figure 12.
(a) and (b) The real (M') and imaginary (M'') parts of electric modulus as a function of frequency at room temperature for $Mn_{1-x}Bi_xFe_2O_4$ *(where x = 0.0, 0.05, 0.1, 0.15, and 0.2) nanoparticles.*

x = 0.0, 0.05, 0.1, 0.15, and 0.2) nanoparticles, respectively. M′ approaches zero in lower-frequency district and shows constant scattering moving in midfrequency area, while it displays a level sort conduct in higher frequency. This sort of conduct is emerged from the unimportant electrode polarization in lower frequencies and short-range and long-range versatility of charge transporters contributing for the conduction wonders in midfrequencies and recurrence-free electrical conduction in higher frequencies, individually [27]. Also, the focus increment M′′ shows the unwinding tops from lower-frequency area to higher-frequency area. The relaxation tops moving plainly demonstrate the spread of relaxation time of the material. The presence of relaxation tops at lower-, center-, and higher-frequency areas is ascribed to the commitment of capacitance to the grain limit, structural, and grain stage attributes, separately, which is very much concurred with the revealed writing.

3.3.7 Nyquist plot

Figure 13 gives out the variation of complex electric modulus spectrum (Nyquist plot) (M′ vs. M″) at room temperature for $Mn_{1-x}Bi_xFe_2O_4$ (where x = 0.0, 0.05, 0.1, 0.15, and 0.2) nanoparticles, and it shows the single semicircular segment for all samples, which is beginning from the grain limit commitments of the material. The Nyquist plot reveals the presence of a non-Debye type of relaxation phenomenon, which is further validated from the electrical modulus analysis [28]. Also, each crescent bend is almost overlapped with each other and no variety is observed, which calls attention to the nearness of electrical unwinding process in the material.

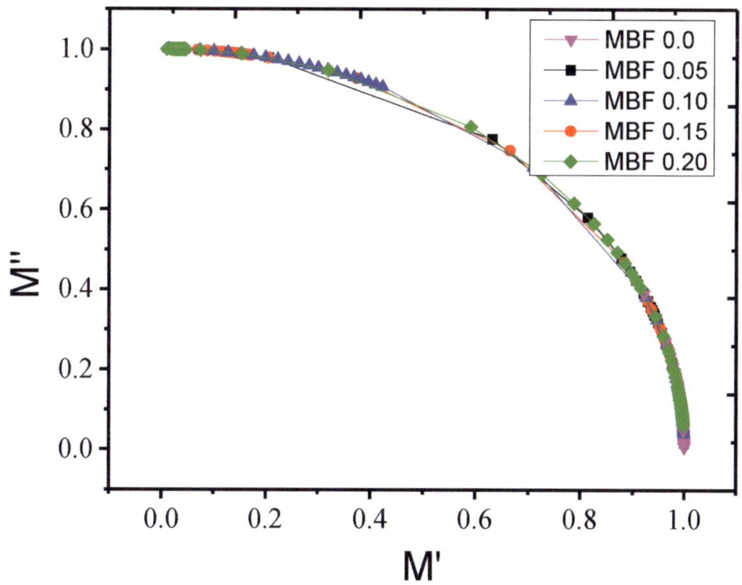

Figure 13.
Nyquist plot of $Mn_{1-x}Bi_xFe_2O_4$ (where x = 0.0, 0.05, 0.1, 0.15, and 0.2) nanoparticles at room temperature.

3.4 Magnetic properties

At room temperature, magnetic hysteresis (M-H) curves of $Mn_{1-x}Bi_xFe_2O_4$ (where x = 0.0, 0.05, 0.1, 0.15, and 0.2) nanoparticles show very narrow hysteresis loops, which confirms the soft magnetic material nature as depicted in **Figure 14**. The magnetization measurement was done by using VSM mounted with maximum applied field of 2 T at room temperature.

The remanent magnetization (M_r) and M_s was estimated by the M-H loop of the y-axis (magnetization axis). The coercivity field (H_c) was estimated by the M-H loop of the x-axis (field axis). The remanence ratio (S) was calculated by using equation $S = M_r/M_s$. The magneton number (η_B) was calculated by using equation $\eta_B = \frac{M.W X M_s}{5585}$. The cubic anisotropy ($K_c$) was calculated by using the equation $K_c = \frac{H_c X M_s}{0.64}$. The uniaxial anisotropy (K_u) was calculated by using the equation $K_u = \frac{H_c X M_s}{0.985}$ [8, 29]. At room temperature, the magnetic parameters such as M_r, M_s, H_c, S, η_B, K_c, and K_u decrease with the increase of Bi^{3+} concentration (**Figures 15–17**). In ferrites, the coercive force is gotten by inversion of the direction of the divider development and that of domain turning around the bearing of the applied field. Generally, the successful sticking for domain wall causes the coercivity; it is understood that the greater grain size reduces H_C [30]. In the present examination, the coercive values

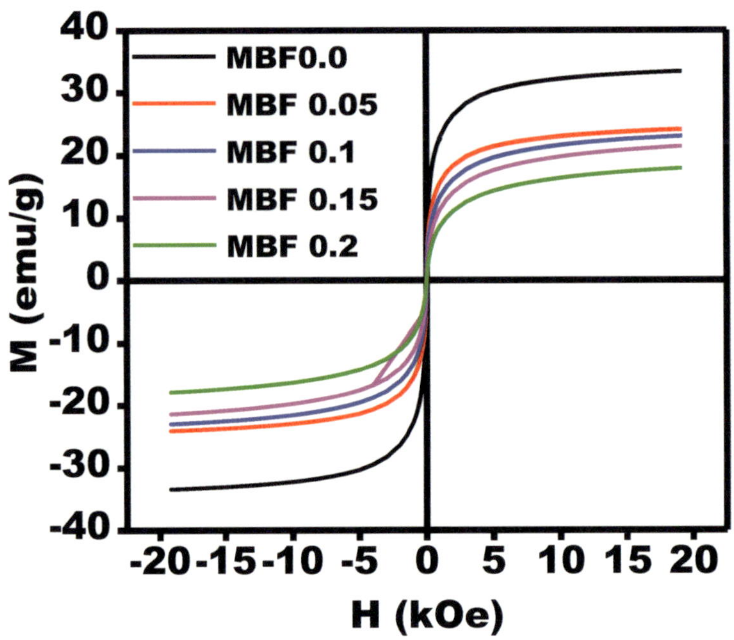

Figure 14.
M-H loops measured at room temperature for $Mn_{1-x}Bi_xFe_2O_4$ (MBF) ferrite nanoparticles with x = 0.0, 0.05, 0.1, 0.15, and 0.2 nanoparticles.

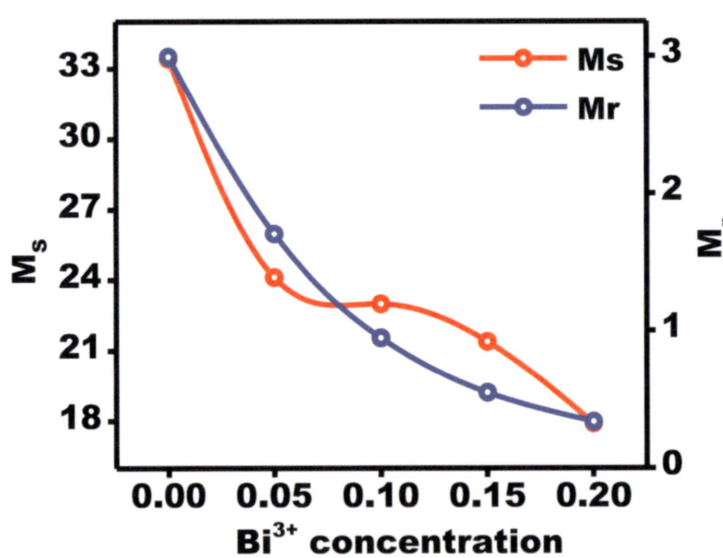

Figure 15.
Variation of M_S and M_r of $Mn_{1-x}Bi_xFe_2O_4$ with x = 0.0, 0.05, 0.1, 0.15, and 0.2 ferrite nanoparticles with Bi^{3+} doping concentration.

are low; thus, the likelihood of domain rotation is lower. The materials with bigger grain size have been used to accomplish lower core loss. Strikingly, as anticipated, the estimation of the reduced remanence value was found to be the lowest for x = 0.20 concentration sample (**Table 4**). We observed S esteems began interestingly from the higher super paramagnetic and the monotonic decrease in S with

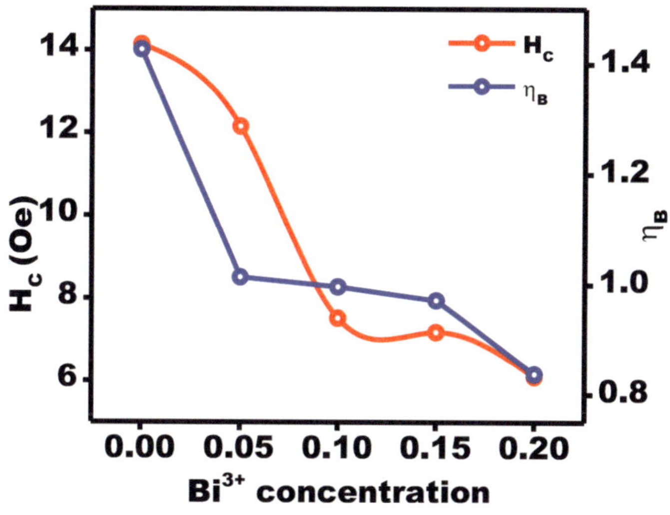

Figure 16.
Variation of H_C and η_B of $Mn_{1-x}Bi_xFe_2O_4$ with x = 0.0, 0.05, 0.1, 0.15, and 0.2 ferrite nanoparticles with Bi^{3+} doping concentration.

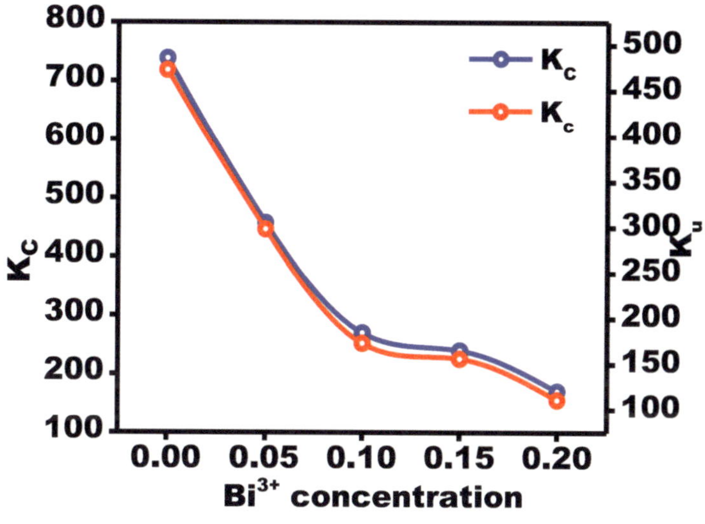

Figure 17.
Variation of K_C and K_u of $Mn_{1-x}Bi_xFe_2O_4$ with x = 0.0, 0.05, 0.1, 0.15, and 0.2 ferrite nanoparticles with Bi^{3+} doping concentration.

Bi^{3+} concentration (x)	M_S (emu/g)	M_r (emu/g)	H_c (Oe)	$S = M_r/M_S$	K_u (erg/ cm^3)	K_c (erg/ cm^3)	η_B (μ_B)
0.00	33.42	2.994	14.14	0.0896	472.55	738.37	1.4259
0.05	24.11	1.705	12.15	0.0707	297.93	457.71	1.0123
0.10	22.98	0.947	7.52	0.0412	172.80	270.02	0.9959
0.15	21.38	0.548	7.18	0.0256	155.84	239.86	0.9713
0.20	17.89	0.337	6.10	0.0189	110.79	170.49	0.8373

Table 4.
Room temperature magnetic parameters of $Mn_{1-x}Bi_xFe_2O_4$ (where x = 0.0, 0.05, 0.1, 0.15, and 0.2), i.e., M_S, M_r, reduced S, (H_c), K_u, cubic anisotropy constant (K_c), and η_B.

increase in Bi^{3+} concentration. Thus, it is expected that the observed impact is overwhelmed by the surface pinning and the change in intrinsic magnetic anisotropy of the particles because of Bi doping.

4. Conclusion

The $Mn_{1-x}Bi_xFe_2O_4$ (where x = 0.0, 0.05, 0.1, 0.15, and 0.2) nanoparticles are synthesized by the solution combustion method using a mixture of fuels as carbamide and glucose. The crystallite sizes were found in nanometer, and the crystallite size decreases with the increase of Bi^{3+} concentration. The lattice parameter increases with the increase of Bi^{3+} concentration due to the difference in ionic radius of Mn^{2+} and Bi^{3+} ions. The porous nature of the samples was confirmed by scanning electron microscopy. The elemental composition was analyzed by EDS composition. The dielectric parameters such as real part and imaginary parts of dielectric constant and dielectric loss tangent decrease with the increase of frequency, and this behavior could be explained by a Maxwell-Wagner interfacial-type polarization. All dielectric parameters increase with the increase of Bi^{3+} concentration. The AC conductivity increases with the increase of frequency according to the hopping model. The real part of complex impedance decreases with the increase of frequency. The observed relaxation peak in the magnitude of imaginary part of complex impedance Z'' demonstrated the existence of space charge relaxation. The Cole-Cole plot reveals one semicircle was obtained for each of the samples, due to the grain boundary effect. It is attributed to the presence of a non-Debye type of relaxation phenomenon. The real and imaginary parts of electric modulus vary with frequency. The Nyquist plot reveals the presence of a non-Debye type of relaxation phenomenon, which is further validated from the electrical modulus analysis. The magnetic hysteresis curves of all samples reveal the soft magnetic material nature. The magnetic parameters such as remanent magnetization, saturation magnetization, and coercivity field were decreased with the increase of Bi^{3+} concentration. $Mn_{1-x}Bi_xFe_2O_4$ (where x = 0.0, 0.05, 0.1, 0.15, and 0.2) nanoparticles have great potential for photonic applications.

Author details

V. Jagadeesha Angadi[1*], H.R. Lakshmiprasanna[2] and K. Manjunatha[2]

1 Department of Physics, P.C. Jabin Science College, Hubballi, India

2 Department of Physics, School of Engineering, Presidency University, Bengaluru, India

*Address all correspondence to: jagadeeshbub@gmail.com

References

[1] Amighian J, Mozaffari M, Nasr B. Preparation of nano-sized manganese ferrite ($MnFe_2O_4$) via coprecipitation method. Physica Status Solidi. 2006;**3**:3188-3192

[2] Chen G, Wang J, Zhou L, Ma W, Zhang D, Ren F, et al. A facile solvothermal synthesis and magnetic properties of $MnFe_2O_4$ spheres with tunable sizes. Journal of the American Ceramic Society. 2012;**95**:3569-3576

[3] Hu W, Qin N, Wu G, Lin Y, Li S, Bao D. Opportunity of spinel ferrite materials in nonvolatile memory device applications based on their resistive switching performances. Journal of American Ceramic Society. 2012;**134**:14658

[4] Ji R, Cao C, Chen Z, Zhai H, Bai J. Solvothermal synthesis of $Co_xFe_{3-x}O_4$ spheres and their microwave absorption properties. Journal of Materials Chemistry C. 2014;**2**:5944

[5] Lee N, Hyeon T. Chemical Society Reviews. 2012;**41**:2575

[6] Arteaga-Cardona F, Pal U, Alonso JM, de la Presa P, Mendoza-Álvarez M-E, Salazar-Kuri U, et al. Tuning magnetic and structural properties of $MnFe_2O_4$ nanostructures by systematic introduction of transition metal ions M^{2+} (M=Zn, Fe, Ni, Co). Journal of Magnetism and Magnetic Materials. 2019;**490**:165496

[7] Li B, Cao H, Shao J, Qub M, Warner JH. Superparamagnetic Fe_3O_4 nanocrystals@graphene composites for energy storage devices. Journal of Materials Chemistry. 2011;**21**:5069

[8] Jagadeesha Angadi V, Anupama AV, Kumar R, Matteppanavar S, Rudraswamy B, Sahoo B. Observation of enhanced magnetic pinning in Sm^{3+} substituted nanocrystalline MnZn ferrites prepared by propellant chemistry route. Journal of Alloys and Compounds. 2016;**682**:263-274

[9] Zhang CF, Zhong XC, Yu HY, Liu ZW, Zeng DC. Effects of cobalt doping on the microstructure and magnetic properties of Mn-Zn ferrites prepared by the co-precipitation method. Physica B. Physics of Condensed Matter. 2009;**404**:2327-2331

[10] Lakshmiprasanna HR, Jagadeesha Angadi V, Babu BR, Pasha M, Manjunatha K, Matteppanavar S. Effect of Pr^{3+}-doping on the structural, elastic and magnetic properties of Mn-Zn ferrite nanoparticles prepared by solution combustion synthesis method. Chemical Data Collections. 2019;**24**:100273

[11] Srinivasamurthy KM, Manjunatha K, Sitalo EI, Kubrin SP, Sathish IC, Matteppanavar S, et al. Effect of Ce^{3+} substitution on the structural, morphological, dielectric, and impedance spectroscopic studies of Co–Ni ferrites for automotive applications. Indian Journal of Physics. 2019;**65**:1-12

[12] Arabi H, Moghadam NK. Nanostructure and magnetic properties of magnesium ferrite thin films deposited on glass substrate by spray pyrolysis. Journal of Magnetism and Magnetic Materials. 2013;**335**:144-148

[13] Godbole RV, Rao P, Alegaonkar PS, Bhagwat S. Influence of fuel to oxidizer ratio on LPG sensing performance of $MgFe_2O_4$ nanoparticles. Materials Chemistry and Physics. 2015;**161**:135-141

[14] Manjunatha K, Sathish IC, Kubrin SP, Kozakov AT, Lastovina TA, Nikolskii AV, et al. X-ray photoelectron spectroscopy and low temperature Mössbauer study of Ce^{3+} substituted $MnFe_2O_4$. Journal of Materials Science: Materials in Electronics. 2019;**30**:10162-10171

[15] Dutta DP, Manjanna J, Tyagi AK. Magnetic properties of sonochemically synthesized $CoCr_2O_4$ nanoparticles. Journal of Applied Physics. 2009;**106**:043915

[16] Durgesh K, Nemkovski K, Su Y, Rath C. Enhancement of Curie- and spin-spiral temperatures with doping Fe in multiferroic $CoCr_2O_4$ nanoparticles. Journal of Magnetism and Magnetic Materials. 2019;**488**:165378

[17] Younis M, Saleem M, Atiq S, Naseem S. Magnetic phase transition and magneto-dielectric analysis of spinel chromites: MCr_2O_4 (M = Fe, Co and Ni). Ceramics International. 2018;**44**:10229-10235

[18] Manjunatha K, Srininivasamurthy KM, Naveen CS, Ravikiran YT, Sitalo EI, Kubrin SP, et al. Observation of enhanced humidity sensing performance and structure, dielectric, optical and DC conductivity studies of scandium doped cobalt chromate. Journal of Materials Science: Materials in Electronics. 2019;**30**:17202-17217

[19] Manjunatha K, Jagadeesha Angadi V, Srinivasamurthy KM, Matteppanavar S, Pattar VK, Pasha UM. Exploring the structural, dielectric and magnetic properties of 5 Mol% Bi^{3+}-substituted $CoCr_2O_4$ nanoparticles. Journal of Superconductivity and Novel Magnetism. 2020;**71**:1-12

[20] Sathisha IC, Manjunatha K, Jagadeesha Angadi V, Chethan B, Ravikiran YT, Vinayaka K. et al. Enhanced humidity sensing response in Eu^{3+}-doped iron-rich Fe_2O_4: A detailed study of structural, microstructural, sensing, and dielectric properties. IntechOpen [Online First]; 2020. DOI: 10.5772/intechopen.90880.

[21] Koops CG. On the dispersion of resistivity and dielectric constant of some semiconductors at audio frequencies. Physical Review. 1951;**83**:121-124

[22] Jagdeesha Angadi V, Choudhury L, Sadhana K, Liu H-L, Sandhya R, Matteppanavar S, et al. Structural, electrical and magnetic properties of Sc^{3+} doped Mn-Zn ferrite nanoparticles. Journal of Magnetism and Magnetic Materials. 2017;**424**:1-11

[23] Jagadeesha Angadi V, Rudraswamy B, Sadhana K, Ramana Murthy S, Praveena K. Effect of Sm^{3+} - Gd^{3+} on structural, electrical and magnetic properties of MneZn ferrites synthesized via combustion route. Journal of Alloys and Compounds. 2016;**656**:5-12

[24] Kakade SG, Kambale RC, Kolekar YD, Ramana CV. Dielectric, electrical transport and magnetic properties of Er^{3+} substituted nanocrystalline cobalt ferrite. Journal of Physics and Chemistry of Solids. 2016;**98**:20-27

[25] Mangalaraja RV, Ananthakumar S, Manohar P, Gnanam FD. Magnetic, electrical and dielectric behaviour of $Ni_{0.8}Zn_{0.2}Fe_2O_4$ prepared through flash combustion technique. Journal of Magnetism and Magnetic Materials. 2002;**253**:56

[26] Tsay C-Y, Lin Y-H, Wang Y-M, Chang H-Y, Lei C-M, Jen S-U. Electrical transport properties of $CoMn_{0.2-x}GaxFe_{1.8}O_4$ ferrites using complex impedance spectroscopy. AIP Advances. 2016;**6**:055909

[27] Badapanda T, Sarangi S, Parida S, Behera B, Ojha B, Anwar S. Frequency and temperature dependence dielectric study of strontium modified barium zirconium titanate ceramics obtained by mechanochemical synthesis. Journal of Materials Science: Materials in Electronics. 2015;**26**:3069

[28] Sakthisabarimoorthi A, Dhas SAMB, Robert R, Influence

of erbium doping on the electrical behaviour of $CaCu_3Ti_4O_{12}$ ceramics probed by impedance spectroscopy analysis. Jose M. Materials Research Bulletin. 2018;**106**:81-92

[29] Manjunatha K, Jagadeesha Angadi V, Ribeiro RAP, Longo E, Oliveira MC, Bomio MRD, et al. Structural, electronic, vibrational and magnetic properties of Zn^{2+} substituted $MnCr_2O_4$ nanoparticles. Journal of Magnetism and Magnetic Materials. 2020;**502**:166595

[30] JagadeeshaAngadi V, Rudraswamy B, Sadhana K, Praveena K. Structural and magnetic properties of manganese zinc ferrite nanoparticles prepared by solution combustion method using mixture of fuels. Journal of Magnetism and Magnetic Materials. 2016;**409**:111-115